EXPERIMENTAL ENVIRONMENTS FOR COMPUTER VISION AND IMAGE PROCESSING

SERIES IN MACHINE PERCEPTION AND ARTIFICIAL INTELLIGENCE

Vol. 1: Pattern Recognition and Image Analysis: Selected Papers from the IVth Spanish Symposium
(Eds. *N. Pérez de la Blanca, A. Sanfeliu and E. Vidal*)

Vol. 2: Theory and Applications of Image Analysis: Selected Papers from the 7th Scandinavian Conference on Image Analysis
(Eds. *P. Johansen and S. Olsen*)

Vol. 3: Neural Networks in Vision and Pattern Recognition
(Eds. *J. Skrzypek and W. Karplus*)

Vol. 4: Parallel Image Processing
(Eds. *A. Saoudi, M. Nivat and P. S. P. Wang*)

Vol. 5: Advances in Structural and Syntactic Pattern Recognition: Proceedings of the International Workshop on Structural and Syntactic Pattern Recognition
(Ed. *H. Bunke*)

Vol. 6: Active Robot Vision: Camera Heads, Model Based Navigation and Reactive Control
(Eds. *H. I. Christensen, K. W. Bowyer and H. Bunke*)

Forthcoming

Vol. 7: Advances in Pattern Recognition Systems Using Neural Network Technologies
(Eds. *I. Guyon and P. S. P. Wang*)

Vol. 8: Thinning Methodologies for Pattern Recognition
(Eds. *C. Y. Suen and P. S. P. Wang*)

Vol. 9: State of the Art in Digital Mammographic Image Analysis
(Eds. *K. W. Bowyer and S. Astley*)

Vol. 10: Generic Object Recognition Using Form and Function
(Eds. *K. W. Bowyer and L. Stark*)

Series in Machine Perception and Artificial Intelligence – Vol. 11

EXPERIMENTAL ENVIRONMENTS FOR COMPUTER VISION AND IMAGE PROCESSING

Edited by

H I Christensen

Laboratory of Image Analysis
Institute of Electronic Systems
Aalborg University, Denmark

J L Crowley

Institute National Polytechnique de Grenoble
France

World Scientific
Singapore • New Jersey • London • Hong Kong

Published by

World Scientific Publishing Co. Pte. Ltd.
P O Box 128, Farrer Road, Singapore 9128
USA office: Suite 1B, 1060 Main Street, River Edge, NJ 07661
UK office: 73 Lynton Mead, Totteridge, London N20 8DH

EXPERIMENTAL ENVIRONMENTS FOR COMPUTER VISION & IMAGE PROCESSING

ISBN 981-02-1510-X

Printed in Singapore by JBW Printers & Binders Pte. Ltd.

Contents

Editorial - Experimental Environments for Computer Vision & Image Processing

Henrik I Christensen† & James L.Crowley*

† Laboratory of Image Analysis
Aalborg University
Fr. Bajers Vej 7
DK-9220 Aalborg East
Denmark

* L.I.F.I.A.
Institute National Polytechnique de Grenoble
46 Ave Felix Viallet
F- 38036 Grenoble
France

Introduction

Rapid advances in computing technology have made it possible to build computer vision systems that are useful for real world applications. This has driven the discipline advance from image understanding to the design and construction of working vision systems. The concern with working systems has brought about a need for experimental environments both testing individual vision techniques, and for prototyping architectures of vision systems. Such environments can greatly reduce the time and effort required to test vision techniques and to prototype vision systems. In this special issue we have the opportunity to present a diverse cross section of such experimental vision environments.

The construction of a complex system is typically a team effort, where such teams may be a group of people at a single institution or a group of researchers collaborating across institutions. In either case, it is necessary that the team work within the framework of a common process decomposition and interface standards, so that individual efforts can fit together. Experimental environments provide such a framework, by specifying, in advance, processes, data structures and communication protocols. Adopting such an environment assures the group that independently developed components can fit together.

Ideally an experimental environment should provide the following facilities to an end user:

Define standards for interfacing to the environment: The system should provide standards for programming, i.e., how software must be produced in order to be integrated into the system. In addition it should also define homogeneous standards for interaction with the system ranging from data formats to user interfaces. Such definitions are in particular critical for systems which are developed by multiple researchers residing at different geographical positions.

Independence of Operating System: A wide set of different platforms are used both in industry and at research institutions, ranging from MS-DOS based PC's, over UNIX work-stations to super computers like the CM-5™. It is consequently desirable to be able to use the same method / environment at all of these.

Enable recycling of methods: It is often possible to structure the processing so that it may be broken down into a set of 'simple' modules that potentially may be used on other applications. I.e., methods such as edge detection, segmentation, graph matching, visualisation and rendering can often be used in new applications, provided facilities for wrapping of such methods are available.

Support real-time processing: Computer Vision and Image Processing have close relations to applications areas such as robotics. Robotics requires real time processing. In computer vision, the increased interest in active vision implies also that real-time response must be available to accommodate control of the sensory system. It thus desirable to have facilities that allow real-time processing, though in principle only part of the system have to run on hard real-time (control - interaction critical processes).

Provide an abstract interface to hardware devices: Hardware devices such as frame stores and graphics systems are usually equipped with very different facilities for acquisition of images and presentation of results to end-users. To enable construction of portable software, it is desirable to have software that provides an abstraction. I.e., in principle a simple command like grab() is sufficient to provide access to the frame grabber, but to enable tuning of the look-up table, selection of video channels, video formats etc. there is typically an abundance of commands available. A standard interface will, however, simplify training and use of such facilities.

Support debugging of software which uses complex data-structures and multi-processing: Most debugging facilities available today allow browsing of simple data structures like integers, bytes and array structures but due to a limited ability to analyse the used data structures it is usually cumbersome to analyse data provided within a computer vision system. In addition there are typically very limited facilities available for debugging of systems which include multiple concurrent processes. It is, however, of interest to have such facilities in particular when the systems of interest increases in size.

Use of non-homogeneous hardware platforms: It is seldom convenient or efficient to use the same computer architecture for pixel level manipulation, 2-D description, 3-D geometry and symbolic computing. In particular, the image/signal level processing may often be implemented efficiently in fine-grained parallel architectures, while intermediate processing may best be carried out on a limited number of medium size processing elements such as Transputers™ and the symbolic computing is better carried out on a single computer which has facilities for efficient handling of symbolic entities. To accommodate such variety of computer architectures, it is necessary to have an environment that allows interfacing to all the architectures. In particular, it is desirable that the distribution onto such a non-homogeneous architecture is transparent to end-users and that the distribution information may be made available to developers on request.

Methods for carrying out systematic experiments including techniques for simple recording of set-up: To fully characterise algorithms and investigate issues such as robustness, domain of application, efficiency etc. it is desirable to have simple techniques for carrying out and recording results from sequences of tests. Preferably the environment should convenient interface to standard packages for data and pattern analysis.

Interactive facilities for processing and analysis of data at any level in a system: An essential facility during testing and debugging is the ability to monitor and examine both data and the interaction of system components system. It is thus desirable to have standard facilities for visualisation and analysis of run-time data, and if possible facilities for visualisation of the system architecture and the interaction between different system components.

Availability of a large set of standard methods: To support rapid prototyping it is desirable to have access to a wide set of standard techniques as soon as the environment is installed as this will simplify introduction of the system and allow for fast adaptation. In addition experiments should include comparative studies and re-implementation of competing techniques is a time consuming task which typically is of little merit.

It is not presently possible to obtain an experimental environment that accommodates all of the requirements that have been stated above. However, over the last few years, computer vision researchers have developed different environments to provide partial support for these issues.

In this volume we have tried to compile a wide variety of different experimental environments. The objectives have been two fold:

1) To provide an overview of state of the art with respect to experimental environments; and

2) To provide an in-depth description of a wide set of environments to enable prospective users to select a tool appropriate for their area of application.

The description includes tools that are applicable for both teaching and research within computer vision and image processing.

Scanning the volume

The papers in this volume may be grouped into three categories:

1) General purpose experimental environments which includes a wide set of methods;

2) Application specific environments; in particular systems related to robotics; and

3) Basic architectures and facilities for construction of computer vision systems.

General Purpose Systems

In the category of general purpose systems the most widely used system is probably the Khoros system, which is described in chapter 1. That system has been developed at University of New Mexico. The environments which has been available for more than 5 years is a rather complete environment for doing experiments in image processing and analysis. The environment includes a large range of methods (in excess of 200 methods are readily available both in software libraries and as stand-alone modules). The environment includes also a visual interface that enables simple inter-connection of modules to complete systems. In this environment. In addition, it is possible to define simple control structures for doing experiments and repetitive processing. To facilitate implementation of new techniques a wide set of tools for design of software, user interfaces are also available. The system is available in public domain, though its development is governed by the Khoros Consortium (a joint venture between research institutions and commercial companies).

In chapter 2 general purpose environment "Advanced Visualisation System" (AVS) available through the company AVS Inc. is presented. Initially the AVS system was available from the company Stardent as an advanced interface for its line of super-computers. The system is specifically designed for simple construction of applications in terms of an interconnected set of modules. The system was initially designed for processing that lead to high dimensional data structures that subsequently could be visualised. Thus this system contains very impressive tools for visualisation and animation of data. Over the last few years, the AVS system has been ported to a wide range of UNIX work-stations, and its user community has grown larger and more diverse, in particular in the area of image processing and analysis. This has lead to formation of an AVS application center that ports applications or modules to the AVS environment. For example, a majority of the Khoros modules have now been ported. All such modules are available in public domain (presently in excess of 4000 modules are available through not all of them are in the areas of computer vision and image processing). The group at Linköping University has extensive experience with the AVS environment in particular for image processing and visualisation of robotics scenario. In this chapter they describe a range of different application of the AVS environment within these two areas. The chapter gives a good overview of the possibilities available in general visualisation environments.

An alternative to the AVS environment is the Explorer™ product available through Silicon Graphics Inc. (SGI). This system has a high degree of similarity to AVS, though it has not yet been developed to the same degree of sophistication. The Explorer system is delivered free of charge whenever an SGI work-station is acquired, and thus has a rapidly growing user community.

In Norway, a long term effort has been directed at providing a standard image processing and analysis environment to be used at many of the national research institutions. Initially the package was termed "blab", but it has recently been renamed to "XITE". This environment provides an interactive system that allows experimentation with image level processing, as described in chapter 3. The modules which have been developed and maintained by Oslo University, include contributions from a wide range of Nordic institutions. The environments may run on a diverse set of platforms and it includes an extensive set of techniques. These include good facilities for analysis of texture and several dedicated modules for medical image processing. A pipeline-style architecture with sequential processing has been adopted for this package. This tools is also available in public domain.

Another interactive tool which has been designed for use in teaching and research in computer vision is the LaboImage package from University of Geneva, which is described in chapter 4. In this package, a register based approach is adopted. I.e. images or intermediate results are stored in a predefined set of registers. Each of the methods takes input from one or more of the registers and produces output to another register. The package includes a wide range of standard image processing and analysis tools. A major emphasis in the design of this tool has been its suitability for use in teaching and training situations. It contains a large set of comparative methods that allow users to study these methods. Many of the text book examples are present. The system has also been used for a number of industrial studies. Examples of the use of the system for inspection are provided in the chapter.

In the Netherlands, there has been a long and substantial effort in image analysis. The effort has included work on development of a suitable environment to be used both in research and teaching, as describe in chapter 5. A wide range of work-stations have been used in this effort. Thus this package is designed and implemented to run on a variety of different platforms, ranging from low-end Apple Macintosh computers to high-end UNIX work-stations. The environment supports distributed processing across platforms and includes an interactive & visualisation tool which has a certain similarity to the Khoros system. Given the system is based on a long term effort it includes a large collection of methods for image processing and analysis tools, including support for reconstruction of 3-D information. This tool also provides a standard set of rules for interfacing and use which has motivated its use in a number of multi-partner projects. The paper includes a brief description of how it has been used both in national and international projects.

Application Specific Environments

The European Commission's Basic Research programme has funded a consortium to develop techniques for the design and control of continuously operating vision systems. This project, which is named "Vision as Process", brings together six partners from five different countries. The experimental environment described in chapter 6 was developed as a skeleton system to enable different partners to integrate their individual contributions into an operating prototype and to study issues related to system architectures and to control. The SAVA skeleton system is designed to integrate information over time, and provides interpreted control to permit individual modules to adapt their functionality based on symbolic reasoning.

The full system architecture includes modules for practically any level of abstraction in computed vision from image acquisition and processing to symbolic interpretation of the scene. The system has been designed to be distributed over a network of computers, including the possibility for dedicated hardware for image processing. In this paper both a flexible system architecture and a standard module architecture used at all levels in the system is presented.

At Oak Ridge National Laboratory there has been a long term effort in construction of autonomous robots for operational use in a nuclear power plant. The effort includes use of a wide range of sensors including sonar, laser range sensors and traditional video cameras. To process such a large amount of information while ensuring real-time operation it is necessary to employ substantial and dedicated computational resources. The robot used, named Hermies-III, has consequently been equipped with a set of real-time computer boards, a hypercube computer, a UNIX work-station and an Apple Macintosh. To provide a suitable environment that enables integration across all the platforms while ensuring real-time operation the MICA system described in chapter 7 was developed. This system has been used substantially with the Hermies robot and an impressive performance in terms of response time and versatility has been demonstrated. It should be noted that despite the fact that it was developed for use in conjunction with robotics research, it is sufficiently general to be used in computer vision research in general.

The researchers at University of Massachusetts, Amhurst has a long standing tradition of building computer vision systems. Initial work was directed towards construction of knowledge based systems, which later has been complemented by use of such systems in the context of robotics. A characteristic of the effort has been use and integration of a broad range of features to facilitate a robust and efficient processing, analysis and reasoning. To enable this it was at an early stage determined that a suitable basis for processing using such a strategy was use of a common symbolic representation method (termed an Intermediate Symbolic Representation (ISR)). This representation has formed a fundamental basis for both the VISIONS systems, and the commercial product KBVision™ . The representation has for natural reasons evolved over the years to accommodate increasing demands in terms of representational power, versatility and efficiency. In chapter 8 the latest version of the ISR (ISR-3) is described together with a system based on this representation. The system has been specially designed to accommodate use in autonomous mobile navigation as part of the ARPA AUV project.

Basic Architectures for Computer Vision

A dominant problem in computer vision research is the immense amount of labour used for implementation of methods for experimental research, another problem is integration across different sites. To make such work more efficient a Danish effort was initiated five years ago. The aim was design of a Vision Programmer Work-Bench (VIPWOB). Basic requirements was the ability to build systems for all aspects of computer vision, while allowing use of existing software with a minimum or no change at all, and interfacing and use of special hardware. To propitiate these requirements the simple flexible, and yet powerful software architecture described in chapter 9 was devised. The architecture imposes very few restrictions on the programmer, enables use of existing software and supports very broad categories of system architectures. The system has now been in use for a little more than 3 years and it has been used for a wide range of different applications.

The interaction between different computer vision processes can typically be formulated in term of objects with associated features such as inheritance, methods etc. To fully exploit the functionality of object oriented communication it is desirable to have a tool that allow simple communication of such objects between modules. In addition it is desirable to have standard module architectures that impose a standard processing model, but the architecture should be sufficiently flexible to allow introduction of arbitrary methods into the control structure. To enable simple inter-connection of modules which are base don such a system philosophy the Module Network Tool (MNT) has been designed and implemented at the Royal Institute of Technology in Sweden. The tool is sufficiently general to support arbitrary module and system configurations and distribution of such across a set of work-stations, as described in chapter 10.

A basic problem in distribution of image processing techniques across a range of work-stations is maintenance of consistency between such modules. If an object oriented approach is adopted for the programming and/or communication it is further desirable to have simple mechanisms for transfer not only of data but also of associated methods and properties. In the framework of UNIX networking a standard for exchange of simple data types has been developed by Sun Microcomputers. This standard and the associated implementation is called XDR (External Data Representation). To accommodate use of a similar technique, but for object oriented communication an External Object Representation (XOR) has been developed at the Royal Institute of Technology in Sweden. Through use of this methods it is simple to construct and interconnect modules which uses Object Oriented Programming. The system has been designed so that it may be used with the traditional C-programming language, with C++ and the LISP package CLOS. The technique, which is described in chapter 11, is in highly interesting as it in principle may be integrated into any of the earlier described experimental environments.

Summary

A wide set of different systems are presented in this volume. In general they provide a good overview of the techniques available to researchers for performing experimental work. It should be noted that one major effort, the ARPA Image Understanding Environment, is not presented here. Unfortunately, it was not possible to obtain contributions from any of the researchers involved in that effort. It is, however, felt that a comprehensive and adequate presentation is provided. We hope the readers will find the volume timely and suitable companion in their experimental work.

We would like to thank the reviewers for their help in bringing together a set of high quality papers in a minimum of time. The editorial effort has been supported by the CEC DG-III Basic Research Office through the project "Vision as Process-II" (EP-7108 VAP-II) and the Danish Technical Research Council through the "MOBS program", this support is gratefully acknowledged.

The Khoros Application Development Environment

John Rasure and Steven Kubica

Khoral Research Inc., 4212 Courtney NE
Albuquerque, New Mexico, 87108, USA
khoros-request@chama.eece.unm.edu

The Meaning of the Word **KHOROS**

Khoros is not an acronym; rather, it is the greek spelling of the word *chorus*. A definition of a chorus is: *a company of singers or dancers who sing or dance in concert*. The name Khoros is symbolic of the objectives for the Khoros project and software. The *singers or dancers* are the interacting systems of Khoros; the objective is to provide a group of systems that will, when used *in concert*, provide a harmonious environment for its many different kinds of users. These same themes are expressed in the definition: *a composition to be sung by a number of voices in concert*.

A distinctly different definition of chorus, *a company of singers and dancers in Athenian drama participating in or commenting on the action*, is symbolic of the free availability of Khoros and the participation of the user community in the development of Khoros.

1. Introduction

This paper is a summary report of the continuously evolving Khoros system. The developers of Khoros have been active in building experimental environments since 1987 [1-3] and realize that the implementation of such a system is a moving target.

In the past, Khoros has often been identified as an image processing system or a data flow language. This paper attempts to clearly explain and substantiate that it is much more that either of those two things. The long term objective of the Khoros project is to research and develop a complete application development environment that allows a casual user to progress from an experiment to a product with minimal effort. The fundamental challenge is to make the environment as broad as possible while hiding its inherent complexity. Khoros 1.0 began to make inroads towards this objective. Khoros 2.0 was developed after a complete reevaluation of the long term objective; the result was a complete rewrite of the system.

This paper is organized to give the reader a tour of the Khoros system. Khoros is a very broad system, and to give the reader a better conceptual understanding of the entire system and not just aspects of the system, detail is sometimes lacking in this summary report. Section 2 presents a brief retrospective of Khoros, evaluating strengths and weaknesses. A number of other systems are drawn into the discussion and contrasted with Khoros. Section 3 provides an overview of Khoros 2.0, followed by a Section 4 on the details of the environment's infrastructure. Section 5 summarizes the application domains that are most likely to interest readers working in the areas of computer vision, image processing and graphics.

A history of the Khoros project is provided in Section 6 before a **Case Study** is presented. The purpose of the case study, Section 7, is to give an actual example of how Khoros is used as an experimental environment. Section 8 closes the paper with information on the status and availability of the system.

2. Background

Khoros is generally thought of as a quality, commercial product even though it is freely available. The perceived strengths come from a variety of characteristics of the system; good e-mail support, good image processing functionality, easy-to-use, supports collaboration, good tools for quickly prototyping new software, etc. However, many users have identified enhancements that would improve the system and experienced users occasionally find limitations in the system.

This input from the user community was carefully considered during the design and specification phases of Khoros 2.0 development. The literally thousands of comments received were distilled into requirements documents and feature lists. The following paragraphs address several of the highest priority characteristics that an experimental environment should have:

1. Extensibility

   An application builder, development environment, or experimental system must be extensible. Designers of such systems can not anticipate all of the needs of the user. Extensibility can mean many different things; modification of the development environment itself, incorporation of new code or data structures, or customizing the appearance or presentation of some aspect of the environment.

   When Khoros is compared to other environments, such as KBVision or AVS [4], Khoros is the most extensible. Khoros includes more code development tools, and source code of the entire system is provided. But Khoros 1.0 was not without problems. The learning curve required by a new user in order to incorporate new code into the system was too steep. A new user needed to read hundreds of pages of documentation and learn a handful of Khoros and UNIX tools before they could begin code integration. (This issue is addressed in Section 4.2.)

2. Data Management

   Data management must be an integral part of an experimental environment. Khoros 1.0 was weak in this area in several ways; lack of an integrated database and inability to process huge data sets. Many of the other experimental environments have similar limitations.

Khoros 1.0 was limited to processing data files that the UNIX host could manage with virtual memory. Khoros 1.0 running on a typical workstation is limited to about 10 to 20 megabyte data files. Many commercial imaging systems rely on custom hardware or non-portable software to get around this limitation. (This issue is addressed in Section 4.1.)

3. Programming Toolkits

   An experimental environment should provide the user with reusable libraries so that the need to write new, custom code is limited. In the case of Khoros 1.0, the libraries to access data, display images, and render graphics were adequate but lacked robustness and consistency. The *object oriented* view of software development provides a methodology for addressing these shortcomings. (This issue is addressed in Section 4.2.)

4. Generality

   As an experimental environment becomes more general, it may lose ease-of-use or it may lack functionality in specific application domains. There was concern from the Khoros user community that the broad focus on software development would leave behind the image processing functionality.

   On the other hand, others in the Khoros user community were being held back by the lack of geometry visualization and data structures that supported computer vision. These users demanded broader functionality. (This issue is addressed in Section 5.)

5. Interoperability

   The term interoperability, like extensibility, means many different things; importing and exporting of data should be transparent, standard interprocess communication (IPC) should be used, standard protocols and programming interfaces should be used, etc. What this means to the user is that other software that they use should be integrated into the environment for them, or easy to integrate by them. Khoros 1.0 has been integrated with software systems such as SAS, S-Plus, IMSL, and MATLAB by the user community. Additional interoperability is desired by the user community, but often companies are not willing to document and share protocols and data structures.

3. What Is Khoros?

Khoros is a software integration and development environment that emphasizes information processing and data exploration. The interacting systems of Khoros 2.0 meet the software development needs of a corporate work group or a collaborative scientific effort by supporting four different levels of user interaction or programming.

3.1. The User's View Of Khoros

The following levels of interaction or programming are available to a Khoros user. An individual will often cross the user levels depending on their experience and the project's needs.

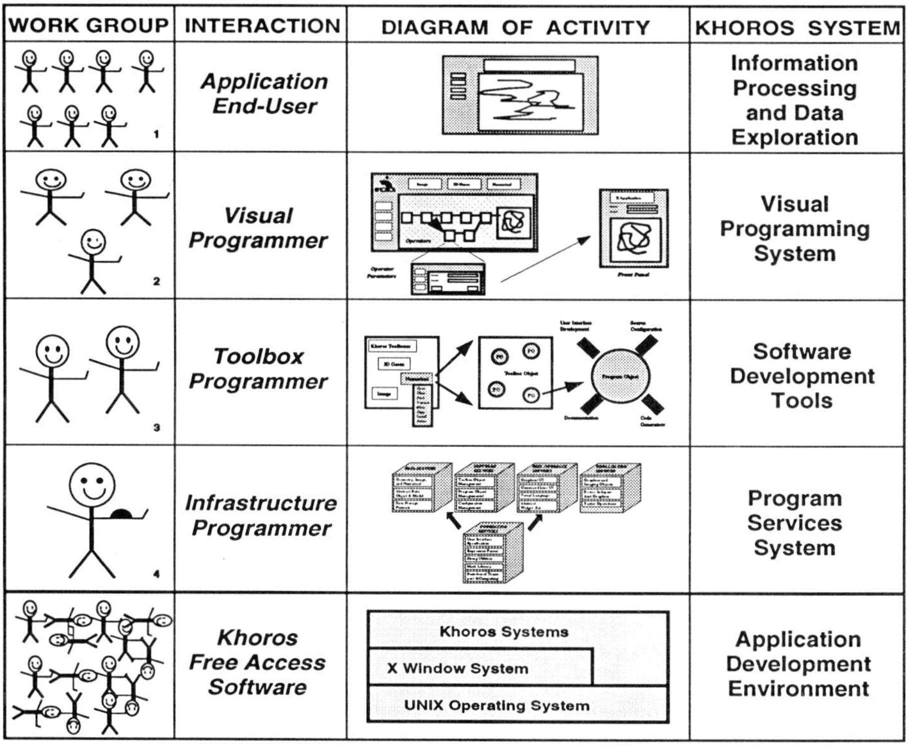

Figure 1: The levels of programming or interaction between the user and the Khoros Systems.

Level 1 An *application end-user* is a person performing production work utilizing one of the Khoros interactive, stand-alone, **information processing and data visualization** applications such as editimage (image display and manipulation), or xprism (2D and 3D graphics, plotting, and data visualization). Alternatively, the same person might use a custom application developed by a coworker who is programming at one of the three interaction levels described below, see Figure 1.

Level 2 Stand-alone applications are often limited in generality because of the required trade off made to deliver ease-of-use. Visual programming supports the interactive development of complex applications by allowing data operators and visualization operators to be combined into a new application. Data operators are non-interactive programs for data manipulation, image processing, geometry visualization, and signal processing. Visualization operators are interactive programs for plotting, rendering, and displaying data. Both kinds of operators are available in the visual programming language. Since there are hundreds of operators to choose from, they are grouped together by function into **toolboxes.** The *visual programmer* accesses the operators from the toolboxes via menus and icons and then uses them to draw, or program, an application at a very abstract level.

The visual programmer can encapsulate a data flow visual program into a stand-alone application for the level 1 application end-user by utilizing an integrated, interactive Graphical User Interface (GUI) design tool. The GUI design tool allows the programmer to create a custom user interface for the visual program. Data flow visual programming and GUI building are used as an integrated **visual programming system** to develop an application that can be delivered as a self-contained information processing and data visualization system.

Level 3 The visual programming system is only as strong as the operators or programs that are provided with the toolboxes, so, at some point, there is a need to either integrate existing software into Khoros or to extend the capabilities of Khoros by developing new software. The Khoros **software development tools** support the process of developing a new toolbox by providing graphical user interface design tools, code generators, documentation support, and utilities for source configuration and management. These computer aided software engineering (CASE) tools act as the *toolbox programmer*'s assistant. The toolbox programmer also uses level 4 as a source of reusable code for the development of custom software.

Level 4 The Khoros **program services system**, level 4, provides high level programming libraries that overlay UNIX and the X Window System (X11R4 and X11R5). These same libraries form the infrastructure of the Khoros system; they are designed to provide a broad-based software development technology for the toolbox programmer. Some of the services that are provided include distributed computing, data management and storage, GUI manipulation and design, and image and graphics visualization.

The program services system is provided in source form so that it can be extended by a Khoros *infrastructure programmer*. The infrastructure programmer extends the Khoros base technology by adding layers or by enhancing existing layers to provide the services required by a new application domain.

Figure 1 summarizes the four levels of Khoros interaction or programming just introduced. Khoros 2.0 is a tremendous improvement over Khoros 1.0 because of the attention given to each level of interaction and programming that a work group requires. The following background sets the stage for a complete discussion of the terminology and concepts discussed above.

The need for a software integration and development environment that emphasizes information processing and data visualization is evident from the explosion of commercial systems which refer to themselves as application development systems. The basic premise is that with the power of the operating system, advanced languages, and windowing systems, a casual user of an application development system should be able to develop domain specific programs very efficiently and deliver them as a quality product. Although this concept is simple and deserves a solution, an elegant and powerful implementation of such a system does not yet exist. The goal of the Khoros project is to research and develop such a system.

The fundamental challenge in developing a complete application development environment is to create a very broad technology base while at the same time hiding its inherent complexity. The application development environment should provide a layer of abstraction above current operating and windowing systems, that dramatically improves the productivity of the software engineer. Major improvements in software technology are being held back due to the instability of the underlying layers. Which operating system should be used? Which windowing system should be used? Which standards should be followed?

3.2. Free Access Software

Khoros is distributed via Internet as **Free Access Software**. Source code and binaries are available throughout the world via Unix File Transfer Protocol (ftp); it can be used by any organization, but it can not distributed for profit without a license. This method of distribution has motivated an estimated 20,000 people to use Khoros, while at the same time retaining the intellectual property rights of the developers. The developers of Khoros believe that this method of distribution is the most aggressive technique available to advance technology rapidly. User participation is critical to the evolution of Khoros and is the basis for the Khoros Consortium, a group of over 25 government and commercial organizations that oversee the technical direction of Khoros.

The objective of the Khoros project is to build a complete application development environment that redefines the software engineering process to include all members of the work group, from the application end-user to the infrastructure programmer, in the productive creation of software. The Khoros project is accomplishing this by providing a broad set of program services that can be used as a source of reusable code for visual program development.

3.3. The Implementation View Of Khoros

The user's view of Khoros is that of a group of four interacting systems, where each system is designed for a specific level of interaction or programming paradigm. A description of the implementation view of Khoros further explains how users interact with Khoros and what capabilities are available.

Khoros 2.0 is implemented as a single **core toolbox** and a set of **application specific toolboxes** as illustrated in Figure 2. The core toolbox contains the **program services system**, level 4, the **software development tools**, level 3, and the **visual programming system**, level 2. The application specific toolboxes contain the information processing and data visualization programs that are used as operators in the visual programming system, level 2. The operators in the application specific toolboxes can also be used standalone by a level 1 **application end-user**. In other words, the Khoros core toolbox

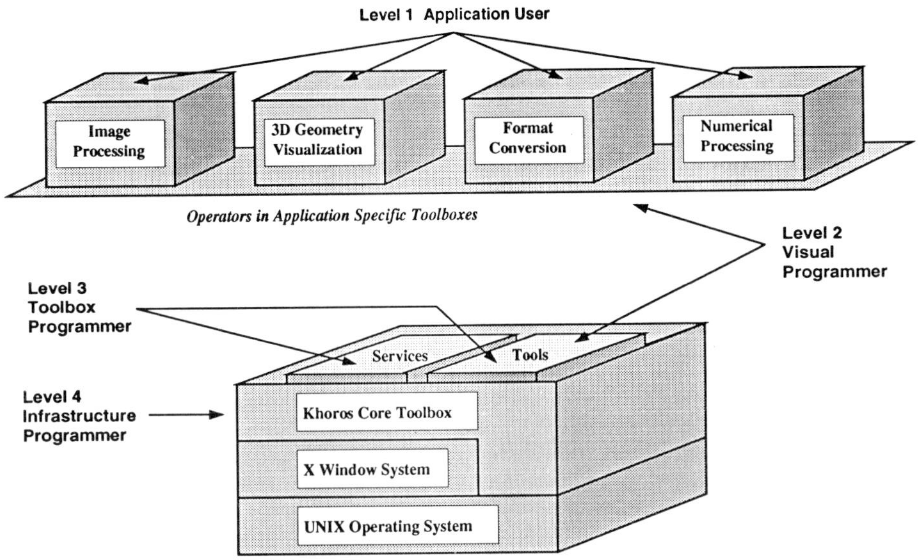

Figure 2: A simplified implementation view of Khoros and its relationship to the user's view, see Figure 1, of Khoros. The Core Toolbox acts as a virtual operating and windowing system. See Table 1 for a listing of the core toolbox programming tools.

comprises a set of libraries, referred to collectively as program services, and programming tools. While the application specific toolboxes contain visual language operators and interactive end-user applications.

The visual programming capability of the core toolbox is given functionality by the operators of the application specific toolboxes. The operators or programs of the application specific toolboxes may utilize the libraries of the program services or they may be implemented completely independent of Khoros. This is an important feature that allows end-user applications to be separated from the rest of the Khoros environment.

The relationships between the user's view and the implementation view illustrates how the Khoros application development environment provides the user with a broad technology base while hiding the complexity of the implementation from the user. A more detailed explanation of the features and capabilities of the Khoros core toolbox is given in the next section.

4. The Khoros Core Toolbox

The core toolbox is composed of program services and programming tools. The programming tools include a graphical user interface design tool, a data flow visual programming language, and CASE tools. Program services, see Figure 3, is organized as groups of libraries that are layered to provide the software developer with a variety of programming interfaces that trade off reduced complexity against detailed control. Foundation services acts as a platform for data services, user interface services, software services, and visualization services; these five services form the infrastructure of Khoros.

4.1. Program Services

The **foundation services** libraries isolate Khoros from the operating system, extend the capabilities of the operating system, and empower the software developer with a broad range of commonly used low-level functions and utilities. The distributed computing library implements both a process abstraction and a transport abstraction that unite a network of machines to make the location of data and compute resources transparent to Khoros applications. This gives the level 3 toolbox programmer the ability to treat the various operating system transports, such as shared memory, sockets or files, as a single *transport object*. Thus, the rest of the Khoros core toolbox is isolated from specific, operating system dependent, transport mechanisms.

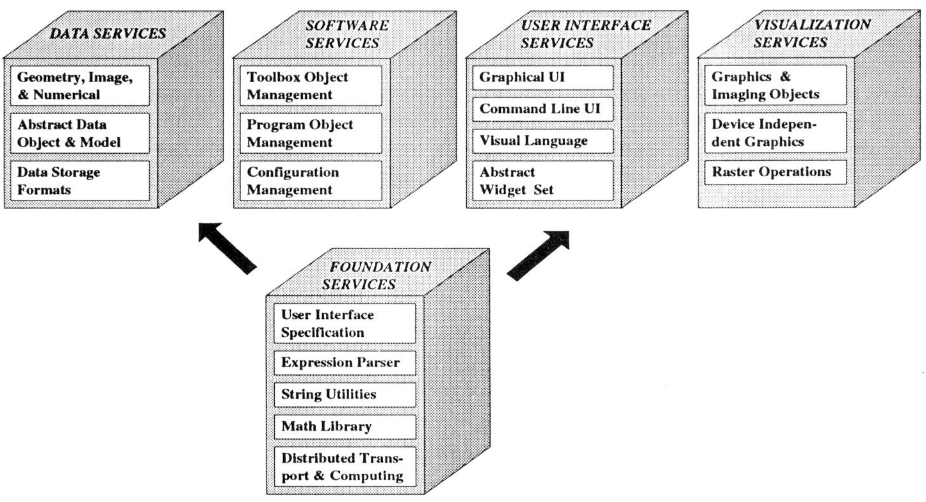

Figure 3: The organization of the program services of the Khoros core toolbox. The libraries in foundation services are utilized by the data, software, user interface, and visualization services.

Command Line User Interfaces (CLUI) and graphical user interfaces (GUI) are both defined by a User Interface Specification (UIS). A UIS file contains a hierarchical and abstract representation of the dialogue between an application's user interface and the user. The UIS library of foundation services maps the textual UIS file into a *form tree* run-time data structure that is utilized by the code generators of **software services** and by the interpreters of **user interface services**. The form tree mediates between the application and its user interface. The UIS file and the form tree separate user interface creation and management from specific user interface implementations.

The foundation services math library contains many commonly used scalar and vector data manipulation functions. The symbolic expression parser and the string parser are used throughout Khoros to support the manipulation of data and the management of text. The foundation services acts to isolate Khoros from operating system utilities, operating system function calls, and any specific user interface system.

Software services is designed to give the toolbox programmer a coherent view of the various components of a program to maximize software development productivity. A hierarchical file system and existing UNIX utilities splinter the development and creation of documentation, source code, configuration management, and user interface specification into disjoint parts and activities. The composition of these parts into a coherent whole is called a *program object*. The program object approach enables more effective automatic documentation generation, code generation, program installation and program maintenance.

Program objects are grouped by function or similarity into toolbox objects. When the visual programmer of level 2 accesses a data or visualization operator, they are actually accessing a program object from a *toolbox object*. The work group's toolbox programmer, level 3, is responsible for developing the appropriate operators for the work group's application end-user and visual programmer, levels 1 and 2 respectively.

Software services is also responsible for both CLUI and GUI code generation. The form tree created from a UIS file by foundation services contains the necessary information to generate application frameworks. The framework consists primarily of a sequence of function calls found in the libraries of user interface services. Toolbox programmers typically don't need to do any custom GUI or CLUI code development because it is automatically created and maintained by software services.

Another important role of software services is the coordination of the programming tools in the Khoros core toolbox. The GUI design tool, the visual language, and the CASE tools communicate by sending and receiving messages and data to and from program objects. The state of all program objects are stored in a simple, distributed, hierarchical object database.

The **user interface services** contains all the functions needed to present, interactively change, and maintain X11 based GUIs. User interface services also includes the library that implements the multi-mode command line user interface. The CLUI is accessed via a UNIX shell in either prompting mode or by the user typing in the entire program name and parameters. As mentioned earlier, both the GUI and CLUI of an application, are created by interpretation or translation of an application's UIS file.

An interactive application is presented to the end-user as *visual objects*, which are an aggregation of *GUI objects* and *graphics objects*. A GUI object is abstracted away from the specifics of the windowing system by the abstract widget library. The abstract widget library provides a powerful mechanism to interactively edit all visual objects, where an object might be an input file browser, a plot axes, or an image. The libraries layered above the abstract widget library manage the GUI as it is running and provide useful utilities for such things as parameter range checking, error pop-ups, and an interactive help facility. The UIS library's form tree is interpreted by the abstract widget library functions to initialize and create the GUI, and then as the user interface changes, the functions maintain the state of the the form tree. This provides the mechanism to save and restore the state of a GUI as a UIS file. GUI objects supported include the Athena, Motif, and OLIT widget sets.

The **visualization services** relies on the abstract widget library to manage image, 2D, and 3D *graphics objects*. The graphics objects are isolated from specific hardware and display languages by the high-level graphics library and geometry services. Visualization services provides utilities for image display and manipulation, two dimensional plotting, three dimensional plotting, surface rendering, and geometry rendering.

Data services implements a powerful and abstract *data object*, which hides the details of data file format, data storage, and data transport. Furthermore, data services provides the mechanisms for network-wide processing of extremely large data sets. The separation of the algorithm from the data through the use of the data services Application Programming Interface (API) allows data services to present data to the toolbox programmer in a flexible manner independent of storage.

Data services is designed from a broad perspective to encompass the processing of data common to application domains including image processing, signal processing, computer vision, neural networks, numerical analysis, surface geometry, volume visualization, and finite element analysis. This is accomplished by providing a hierarchy of data objects, where simpler application specific data objects are all supported by an abstract data object.

Before discussing the programming tools and utility programs provided by the Khoros core toolbox, it is interesting to observe the *common theme* apparent in the implementation of the foundation, user interface, data, software, and visualization services. All five services include data abstractions that isolate and hide the complexity of the underlying layers. The transport object, GUI object, graphics object, data object, and program object are designed to give the toolbox programmer a consistent, simple, and powerful API.

4.2. Programming Tools

To review, Figure 2 illustrates the implementation view of Khoros by showing four application specific toolboxes, the Khoros core toolbox, and their relationships to the four levels of the user's view of Khoros. This figure shows the level 3 toolbox programmer using the various services to create new toolboxes. The figure also shows the level 2 visual programmer using the **programming tools** and existing toolboxes to create new applications.

The programming tools include programs for the visual programmer, important utility programs hidden from the user, and CASE tools for the toolbox programmer. The core toolbox programming tools are summarized in Table 1.

Khoros Core Toolbox Programming Tools and Utilities	
Program Name	**Program Description**
phantomd	phantomd is the distributed computing daemon program utility. It is responsible for Khoros interprocess communication and managing remote data transport.
craftsman	Application specific toolboxes are managed using craftsman. A toolbox object is opened to access its program objects.
composer	Program objects are managed and created using composer. The GUI design tool, guise, code generators, and documentation generators are all invoked by composer.
cantata	cantata is an advanced data flow visual language that supports coarse grain distributed processes and both stream and block data. Its visual hierarchy, iteration, flow control, and expression based parameters make it a powerful simulation and prototyping system. A programmer can create a front-panel to a visual program to encapsulate the complexity of the data flow diagram and deliver it as a production system.
guise	guise is the interactive GUI design tool. guise provides a GUI front end to the capabilities of the abstract widget library. guise operates by interpreting and modifying the User Interface Specification (UIS) or abstract representation of a GUI. At any stage of the development of the GUI, code in the form of an application framework can be generated automatically from the UIS file. Even after the application has been compiled, the programmer is able to edit GUI objects interactively at run-time because all GUI applications in Khoros are layered on top of the abstract widget library just as guise is.
concert	Any application written utilizing GUI services supports the simultaneous interaction of multiple application end-users at geographically distributed sites. The distributed user interface of the cooperative session is managed by the concert program. concert provides *groupware* for the work group.

Table 1: The foundation, software, and user interface services are used to implement the core toolbox programming tools. The table lists the tool names and short descriptions of the programs.

The following sections on the visual programming system and the software development tools provide more details on the programs presented in Table 1.

4.3. Overview Of The Visual Programming System (level 2)

The Khoros visual programming system, level 2, is the integration of a data flow visual language with a GUI design tool. The name of the visual interface to this capability is cantata. It is important to point out that the visual programming system draws from all five of the program services in the Khoros core toolbox and that cantata is primarily a data flow diagram editor. The visual programmer writes a program by connecting operators to form a data flow graph; operators are selected from the application specific toolboxes, see Figures 4 and 5. The visual programmer can encapsulate a data flow graph into a level 1 application, in the form of a *front panel*, using the integrated GUI design tool. If the desired operator is not available, then the programmer can move to level 3 and integrate in or create new operators using the craftsman and composer software development tools discussed in Section B.2.

4.3.1. The cantata Visual Language

The cantata execution subsystem interprets the visual data flow graph dynamically to schedule operators and then dispatch them as processes. A cantata program is shown in Figure 4. The operators are referred to as being coarse grain because they correspond to an entire process, not a code segment or sub-procedure. The user can select data driven or demand driven scheduling of operators depending on the desired responsiveness and their preferred programming style. Once an operator has been scheduled, the dispatcher is responsible for determining the data transport, communication protocol, and process execution mode. The data transport between the operators can be permanent (file or shared memory) or non-permanent (socket or stream). The communication protocol between cantata and the operator can be as simple as just initiating process execution or as general as continuously updating operator parameters as the process continuously executes. The process dispatch mode can be executed *once* or in *continuous* execution. A visual program runs with minimum overhead when demand driven scheduling is combined with continuous dispatching and non-permanent transport. A visual program is easiest to debug when data driven scheduling is combined with execute once dispatching and permanent transport.

Operators are executed locally or remotely to efficiently use a heterogeneous network of machines. cantata utilizes the phantomd to negotiate the remote data transport and spawn processes on remote machines. The visual programmer assigns operators to specific machines interactively to either optimize execution speed or utilize specific hardware. The remote machines do not need full Khoros installations, just a running phantomd.

The application specific domains of image processing or geometry visualization typically process data as blocks; however, the domains of telecommunications and process control process data as a stream. Cantata operators can consume and produce data either in blocks or in streams, where a block is typically thought of as coarse grain data and a stream is typically thought of as fine grain data.

The cantata visual language extends the basic data flow paradigm to make it a more powerful application prototyping or simulation environment. Data and control dependent program flow is provided by control operators such as if/else, while, count, and trigger. Visual procedures are available to the programmer to support the development of hierarchical data flow graphs with hundreds of operators. Operator parameters

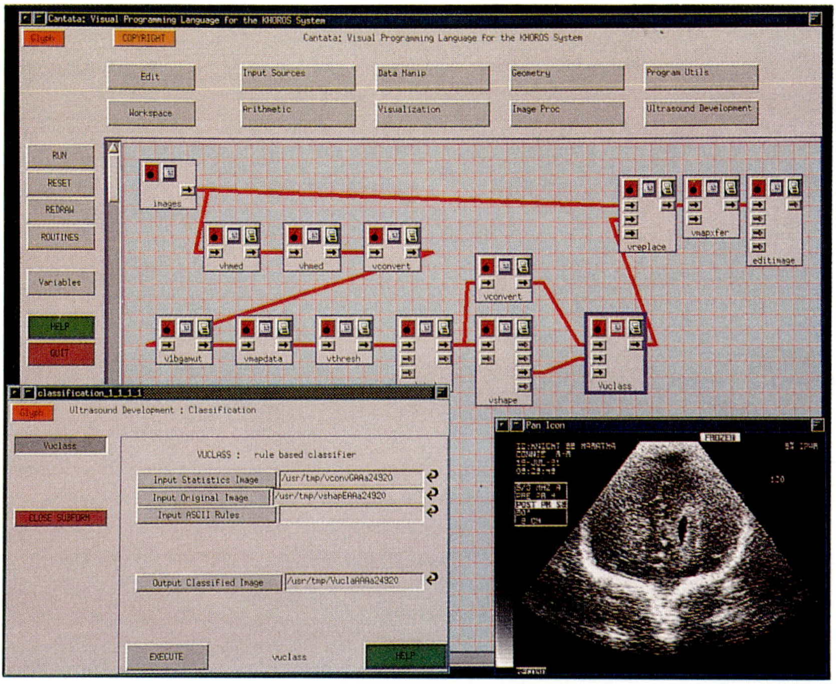

Figure 4: A visual program is written with cantata by interactively drawing a data flow graph. Each node of the graph represents a processing operator, with the lines between the nodes representing the data flow between operators. Control operators and an expression parser extend the functionality of the underlying data flow methodology. Each operator may be "opened" to access a GUI interface from which the operator's parameters may be set.

are set interactively by the user or calculated at run time via mathematical expressions tied to data values or control variables.

It is very important to note that the operators do not need to be compatible with the Khoros data model or linked with any Khoros libraries to be usable in cantata. The connections of the data flow network do not impose any specific data structure or transport. In summary, cantata is used by the level 2 programmer to interactively develop complex, distributed, and interactive applications by combining data operators and visualization operators from level 1 toolboxes into a new level 1 program.

4.3.2. The cantata GUI Design Tool

The visual programmer may use cantata's integrated GUI design tool to *package* a visual program into a production information processing and data visualization

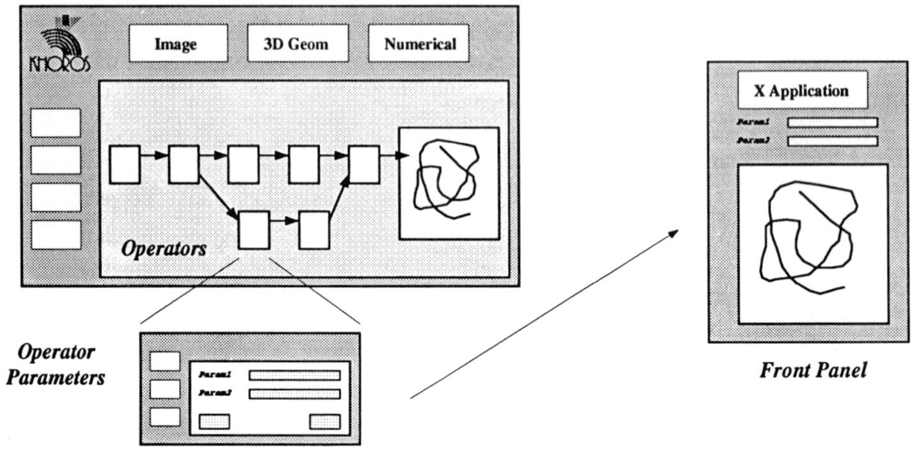

Figure 5: Cantata is used to develop information processing and data visualization applications. Operators are selected from the geometry, signal, or image toolboxes and then placed in the data flow graph. The input and value parameters of the operator are accessed by "opening" the operator icon, or glyph, and then interactively setting them. The visual program on the left is collapsed into a front panel, labeled "X application", by using the integrated, interactive GUI design tool.

application. The visual programmer has full access to all operator parameters, operator connections, and visual language features. This flexibility and accessibility presents too much complexity and requires too much learning for the level 1 end-user if the visual program is to be effective for production work. The GUI design tool can be used to design an alternate, self-contained, GUI for the visual program that allows only limited interaction with operator parameters.

Operator parameters are accessed in cantata by "opening" the operator icon, or **glyph**, and then interactively setting or inputing values to the GUI objects. A visual program with 100 operators may have over 400 GUI objects, all of which affect the results and operation of the program. The visual programmer builds an alternate GUI to the program by selecting the few, most important, GUI objects and then placing them in a **front panel**, see Figure 5. The front panel can be thought of as a simplified, custom user interface for the visual program. Once the front panel is finished, it is given a name so that it can be delivered as a stand-alone application.

It is important to note that the front panel GUI is interpreted; no compiling is needed to run or deliver the front panel as a stand-alone application. Also, the application end-user can select the look-and-feel (Athena, Motif, OLIT) or edit the GUI to better meet their needs without affecting the visual program. The original visual program can always be accessed for further exploration of and more detailed control over the application.

4.4. Overview Of The Software Development Tools (level 3)

A toolbox programmer can use the level 3 software development tools to create operators for the visual language or to develop end-user applications that are completely independent of the visual language. The Khoros software development tools integrate in the access of other commonly used CASE tools to provide a complete software development environment. The craftsman and composer tools support the iterative process of developing, maintaining, delivering, and sharing software in the context of Khoros. They improve productivity by acting as a programmer's assistant; providing automation where possible and enforcing consistency when necessary. In addition, they hide the complexity of the underlying source configuration, code generators, and documentation formatters.

4.4.1. The craftsman Toolbox Manager

A toolbox is a collection of programs that are managed independently from the rest of the work group's software. A work group is likely to be simultaneously creating and maintaining several toolboxes, or they may be preparing a toolbox for delivery. The toolbox programmer manages these activities using craftsman.

A craftsman user performs operations on a toolbox such as, open, close, pack, unpack, delete, move, install, and copy. Figure 6 illustrates an opened toolbox. The toolbox approach to managing software allows the work group to partition responsibilities and eliminate dependencies.

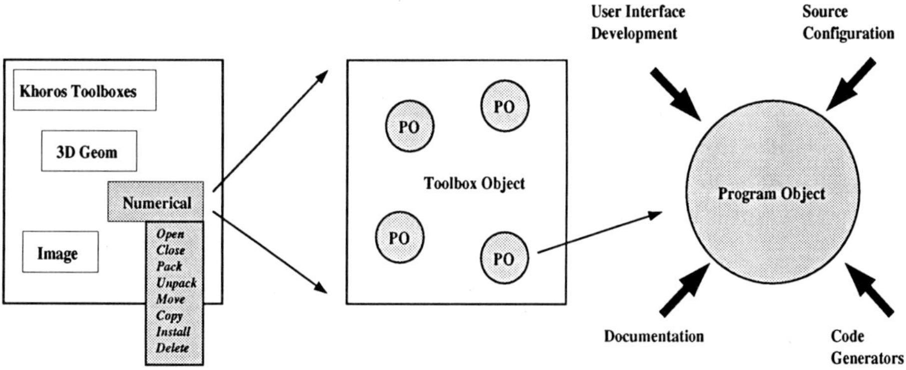

Figure 6: Craftsman and composer are used to develop and manage toolbox objects and program objects. The following sequence proceeds from left to write: 1) the toolbox programmer selects and opens a toolbox icon in craftsman, 2) the opened icon reveals the available program objects, 3) a program object is selected causing composer to be invoked, and 4) the program object's configuration, documentation, code, and user interface are edited using composer.

4.4.2. The Composer Program Manager

A toolbox is made up of one or more program objects, where a program object is the composition of source code, User Interface Specification (UIS), documentation, and configuration files of a single executable operator or application. Each of the program object components requires substantially different kinds of operations; for example, operations on code include generation, compiling, debugging and installation, while operations on a UIS include creation and editing. Composer provides the toolbox programmer convenient access to all of the program object components and can invoke all of the operations needed to create, edit and manage a program object. Furthermore, it manages the dependencies between the components as the toolbox programmer progresses through the software life cycle of a program object.

One of the most important capabilities of composer is the automatic generation of graphical and command line user interface code. The generated code acts as an application framework and eliminates most of the tedious work associated with developing a user interface. Guise is invoked by a toolbox programmer from within composer and then used to interactively design a GUI. The GUI is stored in the form of a UIS that is used as input to the code generators.

4.4.3. The guise GUI Design Tool

The level 3 programmer uses guise to create and interactively design a GUI for either information processing operators or data visualization operators. guise provides the toolbox programmer with a menu of GUI objects that can be interactively placed on a user interface. Once present in the UI, all of the attributes of a GUI object can be directly, interactively edited. An important capability of the guise and code generator combination is that all aspects of the UI are separated from the functionality of the application. Thus, interactive development is supported throughout the lifetime of the application. Figure 7 illustrates guise being used to generate an interface for an information processing operator.

The GUI of a data visualization operator is relatively complex compared to the GUI of an information processing operator. For this reason, the UIS created by guise for the visualization operator is used as input to a GUI code generator to create an application framework. On the other hand, the UIS created by guise for the information processing operator is interpreted at run time. In either case, the UIS is also used as input to a command line user interface code generator.

4.4.4. The concert Groupware System

Concert enables any Khoros application built upon user interface services to operate as *groupware*. The distributed computing capability of foundation services and the user interface event recording capability of user interface services are combined to produce the functionality of the concert program. When concert is used to invoke a data visualization system or operator, or one of the programming tools, it replicates the process to a network of machines and manages the communications between the processes. The result is simultaneous interaction of multiple end-users at geographically distributed

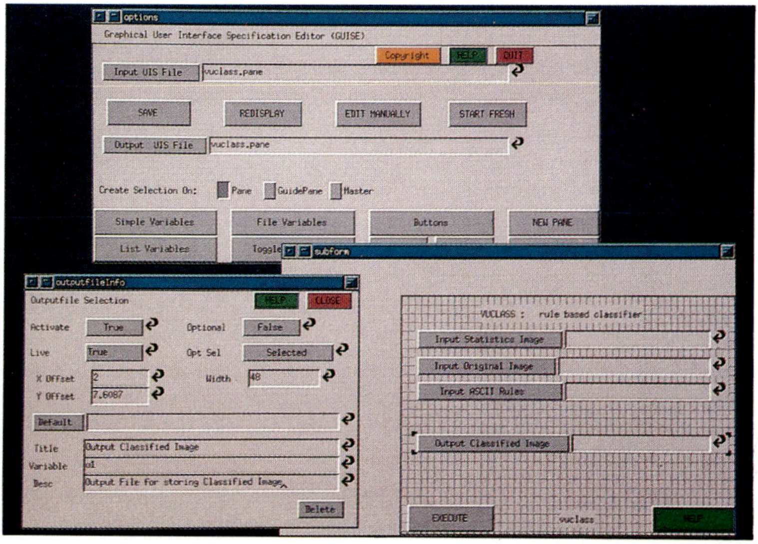

Figure 7: Guise is shown here, upper center, being used for the creation of an interface for a new information processing operator called vuclass. GUI objects are interactively placed on the new interface, lower right. The snap to grid helps to position the GUI objects. Attributes such as size, text, position, default values, and range error checking can be interactive edited at any time during the development of the GUI. The lower left shows the form of attributes of an *outputfile* object. composer is used to configure and install the operator in cantata. Figure 4 shows the GUI being developed in this figure being accessed from within cantata.

sites.

5. The Application Specific Toolboxes

An application specific toolbox is a collection of programs grouped together because of similar function. This document uses the following terms synonymously; programs, program objects, operators, information processing operators, data visualization operators, and stand-alone applications. The *operator term* is used in the context of the visual programmer, level 2, and the *program term* is used in the context of the level 1 end-user or the level 3 user. The term toolbox object is synonymous with the term application specific toolbox.

There are many important aspects of the separation between the Khoros core toolbox and the application specific toolboxes. The fundamental motivation for toolboxes is the need to reduce and manage complexity. Many factors contribute to complexity of a software system; amount of source code (about 500,000 lines of code in the Khoros core toolbox), dependencies between components, dependencies with other systems (UNIX,

X, data storage formats, hardware), and the number of people involved with the software development. Toolboxes provide a methodology for partitioning both a work group and the software into smaller pieces; thus reducing complexity and increasing productivity. The toolbox division also allows for different development schedules.

The toolbox methodology, as supported by craftsman and composer, provides several other important benefits. Collaboration in the use and development of software is enhanced because of the consistencies that are enforced by the tools. The result is that diverse groups can share software and leverage each others resources. Also, portability and maintainability are enhanced because of the embedded source configuration system. Table 2 summarizes the toolboxes being developed by the Khoros group. Figures 8-11 illustrate some different types of visualization which are possible using Khoros. The appendix provides a list of all of the operators in Khoros 2.0.

There are numerous other groups developing toolboxes in application areas such as neural networks, process control, telecommunications, instrument control, medical imaging, image restoration, pattern recognition, morphology, target recognition, finite element analysis, simulation, and modeling.

Khoros Application Specific Toolboxes	
Toolbox Name	**Description**
Image Processing	Image pre-processing, analysis, filtering, feature extraction, segmentation and pattern recognition. The operators in this toolbox are written to the image services API found in data services.
Data Visualization	Display, editing, and animation of images. Display and interaction with 2D, 3D, and geometry data and device independent rendering. Note that this toolbox has a very close relationship to visualization services as presented in Figure 3 and that it utilizes the expression parser to support the *object calculator*.
Geometry Visualization	Display and manipulation of geometry primitives and scenes. Rendering of volumes, iso-surfaces, polygons, spheres, etc. The operators in this toolbox are written to the geometry and image services API's of data services. This toolbox has some overlap with the data visualization toolbox, but is much more complete for scientific visualization.
Format Conversion	Conversion of popular storage formats to import and export data. This toolbox uses the abstract data model of data services as a *virtual format* that can be used to map one format to another.
Data Manipulation	The data manipulation toolbox includes operators or programs that are basic mathematics and algebra. The operators fully support the abstract data model of data services. These baseline operators are used by the image, numerical, and signal toolboxes. The operators in this toolbox are written to the abstract data services API.

Khoros Application Specific Toolboxes	
Toolbox Name	**Description**
Numerical Analysis	Matrix manipulation and algebra (Eigen analysis, linear ordinary differential equations, covariance, LU decomposition). The operators in this toolbox are written to the numerical services API.

Table 2: Summary of the Khoros 2.0 application specific toolboxes.

All of the programs contained in the toolboxes described in Table 2 are written using one of the Application Programming Interfaces (API) provided by data services. Data services includes image services, numerical services, geometry services, and abstract data services. Each service has a corresponding API that mediates between the data model (image, numerical, geometry, abstract) and the data processing algorithm. The image, numerical, and geometry data models are derived from or based on the abstract data model. The result is reusable and polymorphic operators.

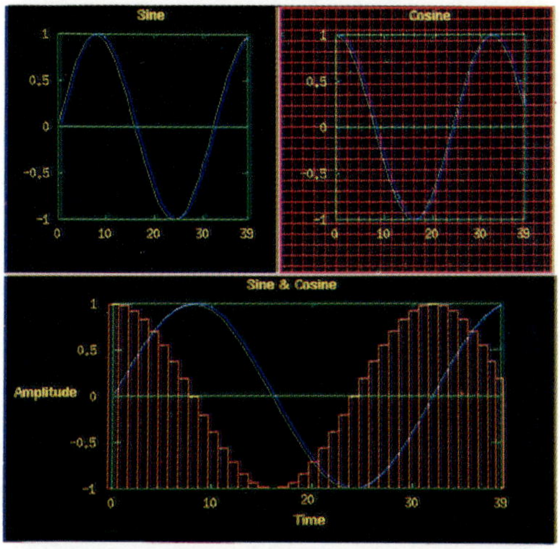

Figure 8: Data plotting is a common and popular type of visualization available with Khoros. Here, the xprism program available in the envision toolbox is used for drawing a number of 2D plots. Multiple plot areas allow for the easy viewing of multiple sets of data in relation to each other.

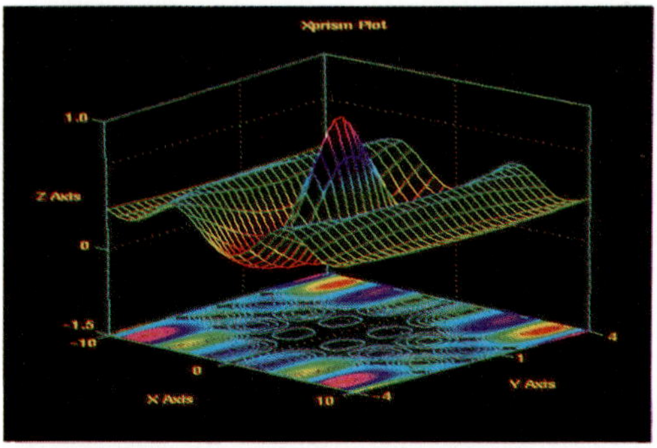

Figure 9: This figure illustrates the use of the xprism program to plot a 3D surface. xprism is a program object in the envision toolbox.

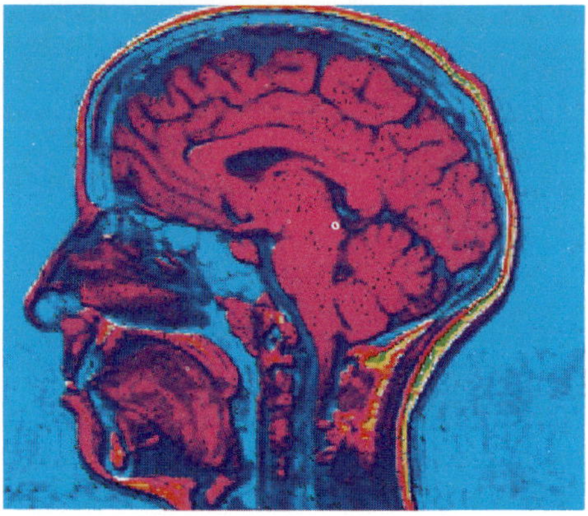

Figure 10: Khoros has become very popular for medical image processing and visualization. Here, the editimage application available in the envision toolbox is used to pseudocolor an image showing an MRI cross-section of a skull.

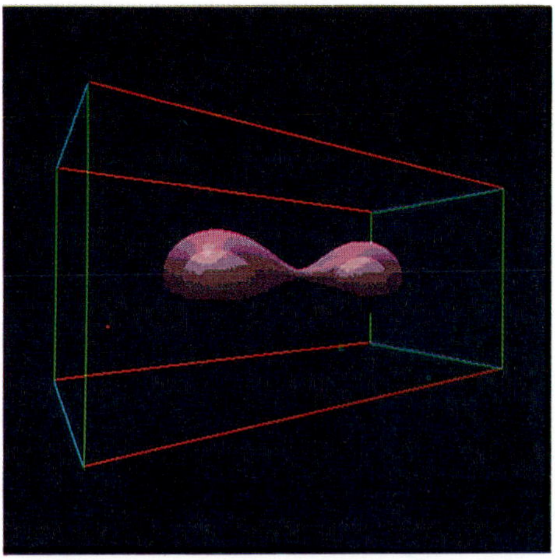

Figure 11: The rendermonster application available in the geometry toolbox is used to visualize concentric iso-surfaces representing different density levels of a ytterbium atom. This application and the geometry toolbox is being developed in conjunction with Wes Bethel at Lawrence Berkeley Laboratory.

The importance of polymorphism should be stressed. There is a huge amount of wasted energy in developing data processing algorithms that do similar things but that can not work together because of differences in data structures and representations. Data services provides a solution for this problem. For example, a one-dimensional signal processing program can operate on any dimension of a time-varying, multi-spectral image or it could even operate on the colors of connected array of polygons. Polymorphism across the different toolboxes is synergistic.

6. History Of Khoros And Short Summary

The scope of the Khoros Project has grown beyond the original effort to provide an effective tool for research in image processing. The project began in 1987 with the creation of XVision 1, see Figure 12. The first step taken was to identify and assemble a collection of fundamental algorithms for image processing. In addition, a 2D/3D plotting package and an interactive image display/editor were written to provide tools for visualizing image data. The goal was to bring together a library of programs with a consistent, easy to use interface to facilitate the quick prototyping of algorithms.

The next major step was the search for a user-friendly interface to the growing library, especially as applications became more sophisticated, drawing on larger numbers of programs in more complicated combinations. Since the results of image processing programs are typically visual, it seemed particularly appropriate that the means of obtaining those results should be visualized also. Therefore, it was natural to look to the exciting new applications of computer graphics, made available with the new generation of less

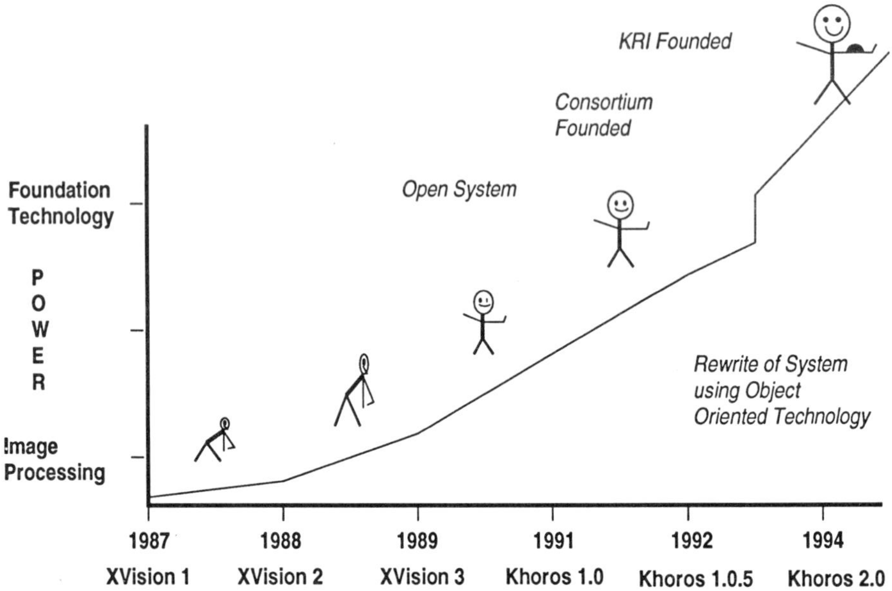

Figure 12: Khoros has evolved from an image processing tool to a broad-based environment for application development and software integration. A major increase in the power of the system occurred with the creation of Khoros 2.0.

expensive, high resolution workstations. The decision was made to base the graphical component on the X Windows System release 10. The result was XVision 2.

Several different graphical user interfaces were considered, tried, and outgrown, with the best features of each retained for the next iteration. What finally emerged was a hybrid approach. A data flow graph, similar to the block diagrams already familiar to practitioners in the field of image/signal processing, was used to access and connect programs from the processing library. A graphical user interface was used to acquire additional information needed by those programs.

As the visual interface to the image processing library was evolving, so were tools to assist in the use of the system. Complete hand coding of programs for the image processing library gave way to the use of code generators. Development and maintenance tools were introduced to support the creation of command line and graphical user interfaces to programs. A high level user interface specification was defined as the foundation for defining the user interface to library functions as well as the tools themselves. These developments were made concrete in XVision 3.

The original simple, pipeline concept of the visual user interface was then extended to a complete, high-level, visual programming language by the addition of control constructs and visual hierarchy.

Meanwhile, other high level application programs were added to the system. At the same time, the image processing library was expanding at a rapid pace, and was joined by a digital signal processing library and an image file format conversion library. The graphics library, the image display library, and the expression parser matured as was necessitated by the needs of the new high level applications that required their services.

These concepts were part of Khoros 1.0, the first public release of the Khoros Group's software. Usage of the system expanded along with the library of algorithms, with new programs continually being added. The development of new programs was always been driven by a specific need. Applications in remote sensing and medical imaging were primary contributors to the Khoros 1.0 library which contains over 250 image/signal processing and pattern recognition algorithms.

The goal of Khoros 2.0 is to cut loose of its domain specific mooring and take flight as a diversified software integration and development environment. This is being achieved by covering a very broad spectrum of application domains, by providing API's for data processing, software development, distributed computing, GUI building, and visualization, and by providing three integrated programming systems. These concepts are perhaps summarized better by Figures 1, 2 and 3. Khoros 2.0 is a base technology that provides information processing and data visualization applications, a visual programming system, software development tools, and a program services system.

7. Case Study : Ultrasound Image Segmentation and Volume Measurement

This case study examines how Khoros was used as a complete software development environment for an ultrasound medical imaging experiment. This final solution required the development of a new image processing operator, as well as a complete end-user application which utilized that operator. The Khoros system was successfully used for each stage of this development [5].

7.1. Problem Description

Blood clots in the brain of a premature infant can prevent the proper drainage of cerebrospinal fluid from the brain ventricles. This may cause the ventricles to enlarge and compress the brain, resulting in brain damage. This damage can be minimized by draining the cerebrospinal fluid before the ventricles become too enlarged; however, the drainage procedure itself presents a danger of infection. So that this procedure is performed only when necessary, diagnostic ultrasound is used to visualize the brains of such infants. Until recently, quantitative measures of the ventricular volume were not possible. It was often up to a physician to visually interpret the relative sizes of the ventricles over a sequence of ultrasound images. The physician would then have to make a *qualitative* judgment as to when intervention was necessary. However, in a research project conducted by Ben S. Brann IV, M.D. at the University of New Mexico Hospital's Department of Pediatrics, a method for determining *quantitatively* at what volume intervention is required was being studied [6].

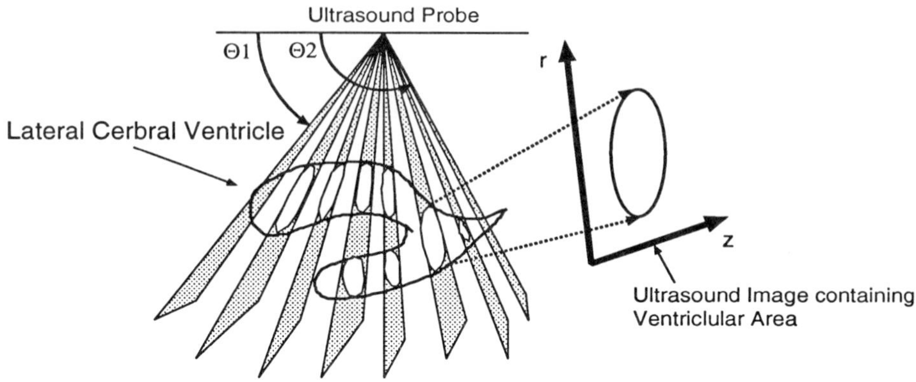

Figure 13: Illustration of mathematical model for calculating ventricular volumes from a radial sequence of ultrasound images. Each radial image plane contains ventricular regions which cover the r and z dimensions. Cylindrical integration over the radial sequence from Q1 to Q2 results in the final ventricular volume.

In these studies, the ventricular volume calculations were performed by integrating the ventricular areas over a radial sequence of ultrasound images. Using this method, the volume of the ventricles could be mathematically calculated within an acceptable degree of accuracy [6]. This model is illustrated in Figure 13. This model was verified using a somewhat cumbersome procedure. Ultrasound images were taken over a radial sequence of positions and digitized. Printouts of these images where then given to a technician who would manually trace out the ventricular regions. A computer program then calculated the volume according to the areas of these manually traced regions. This method, while successful for verification of the mathematical validity of the model, was deemed unsuitable for use in a clinical setting. The time delay between the acquisition of the ultrasound images and the generation of numerical volumes was too long, often taking over an hour, to be of clinical use. What was needed was a reliable, automated software system for image segmentation and volume calculation. The creation of such a system proved to be an excellent opportunity for testing the different levels of interaction with the Khoros system.

7.2. Overview

Creating the software system for the quantitative measurement of ventricular volumes was a two-stage process. The initial stage was to develop an image processing procedure capable of automatically segmenting the ventricular regions from a given sequence of ultrasound images. The next stage was to integrate that procedure into an end-user application which could be used for clinical diagnosis by physicians in the pediatric ward. The end-user application was required to allow the physician to acquire and manage a sequence of ultrasound images, automatically segment those images, and then perform the calculation of the ventricular volume.

The creation of this system spanned many different levels of Khoros interaction. The development of the segmentation procedure was accomplished at the visual programming level of Khoros. Using cantata and the toolbox of image processing operators, a nearly complete procedure was developed. When the existing library of operators was found to be insufficient, a new operator was easily created by moving down to the toolbox programming level. When the segmentation procedure was complete, the final application was developed. The application was also created at the toolbox programming level; it utilized the Khoros services for doing things such as manipulating and displaying the ultrasound images. Note that the application was developed at the toolbox programming level, but the application itself is targeted at the end-user. In this way, the creation of this system covered the first three levels of Khoros interaction, see Figure 1.

7.3. Visual Programming Level

The procedure for automatically segmenting an ultrasound image was developed at the visual programming level. Cantata, in conjunction with the operators available with the image processing toolbox, provide a complete experimental laboratory for investigating potential solutions. Since experiments are so easily prototyped with cantata, it was possible to quickly and effectively iterate through a number of different potential solutions. This iterative approach led to the development of a segmentation procedure which was able to provide reasonable results in an acceptable amount of time. A typical fourteen image sequence could be processed in under ten minutes.

A block diagram of the cantata data flow is shown in Figure 14 and the corresponding cantata program is shown in Figure 4. The first step of this procedure is to preprocess the ultrasound image in order to reduce the level of random noise. This is a standard image processing technique that is accomplished by passing the image through two median histogram filters. The result of this process is a smoother image which helps to improve the performance of the later components of the procedure [7].

The preprocessing is followed by a clustering operator which defines distinct regions within an image. The clustering algorithm uses a heuristic method for partitioning the feature space, which in this case is simply a one dimensional grey level histogram. This space is iteratively divided by bisecting the areas containing the largest population densities [8].

The result produced by the clustering algorithm is a new image that approximates the original spatial distribution of the data with a more compact representation. This new representation contains only essential information needed for the rest of the procedure. Only ventricular and non-ventricular regions with high contrast are preserved. Other

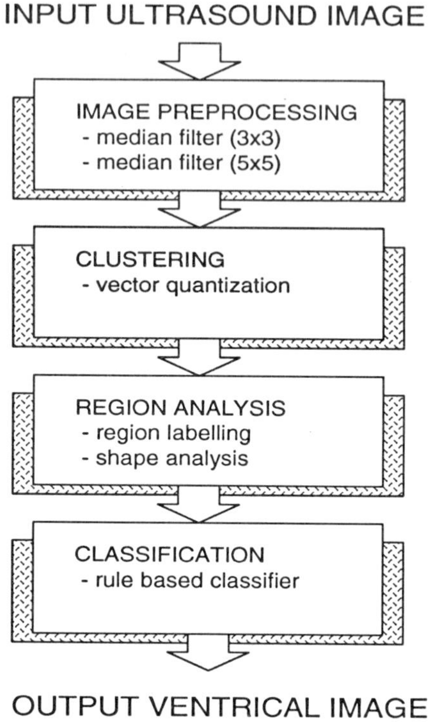

Figure 14: Data Flow of procedure for automatic segmentation of ventricular areas The visual program which implements this procedure is shown in Figure 4.

extraneous information, such as the minor grey-level variations within the regions are removed by the clustering.

Once the clustering is completed, a gray-level thresholding is performed to remove the brighter regions. The ventricular regions are darker than the rest of the ultrasound image, so by thresholding the brighter regions, the obvious non-ventricle regions are quickly discarded. The actual ventricular regions are then a subset of these remaining darker regions.

The process of identifying the ventricular regions is accomplished in three steps. The first step is simply to assign a distinct numeric label to each of the regions so that the other algorithms can identify and distinguish them. The next step is to perform a shape analysis of these labeled regions. The shape analysis step assigns a vector of characteristic information to each region describing its size, shape, and position. The information contained in this vector consists of some standard image processing shape characteristics such as area, x and y variance, eccentricity, overall position, and weighted area [9].

Unfortunately, the final step needed for identifying the ventricular regions could not be accomplished using any of the existing Khoros operators. It was necessary to move down to the toolbox programming level in order to develop a new operator.

7.4. Toolbox Programming Level

Up to this point, standard Khoros image processing operators had been used to identify a number of potential ventricular regions. In order to distinguish the ventricular and non-ventricular regions, the development of a rule-based classifier was necessary. By using the tools and services available at the toolbox programming level, it was possible to quickly develop the new operator.

Composer, used in conjunction with guise, was used to create the vuclass program object. The command line user interface for the vuclass program and the cantata interface for the vuclass operator were automatically generated. The task of handling specific details of data file format, storage, and transport, was also abstracted away since all data access for the ultrasound images was handled with the image services component of application services. The expression parser available in foundation services provided the required capability for evaluating the classification rules. Using the abstractions provided by the Khoros services, the implementation itself consisted only of code which dealt with the functional details of the algorithm. The actual amount of custom code which was written to implement this classifier was under thirty lines.

The classifier compares the characteristics of each region assigned during the shape analysis step to a set of experimentally determined characteristics that are atypical of ventricular regions. The regions that conform to these characteristics are discarded. The remaining regions are then classified as ventricular regions, thus completing the analysis. The rule-based classifier is illustrated in Figure 15.

7.5. End-User Application

With the automatic segmentation procedure completed, the final stage in the development was to build the end-user application. Note that it would have been possible to implement this end-user application at the visual programming level. The visual program for the automatic segmentation procedure could have been created using the "front panel" capability in cantata. This was not done for two reasons. The workstation which was to be used for running the application had an extremely limited amount of disk space. To save on disk space, it was necessary to create a separate application which could be installed without the rest of the Khoros system. Secondly, there is overhead associated with cantata dispatching the operators as processes. This time delay was eliminated by implementing the procedure as a separate application using a series of library calls to the image toolbox library.

The toolbox programming level was used for the development of this application. By using guise, the prototype of the interface was interactively created with help from a physician from the Department of Pediatrics. This interface was used as input to code generators which created the application framework.

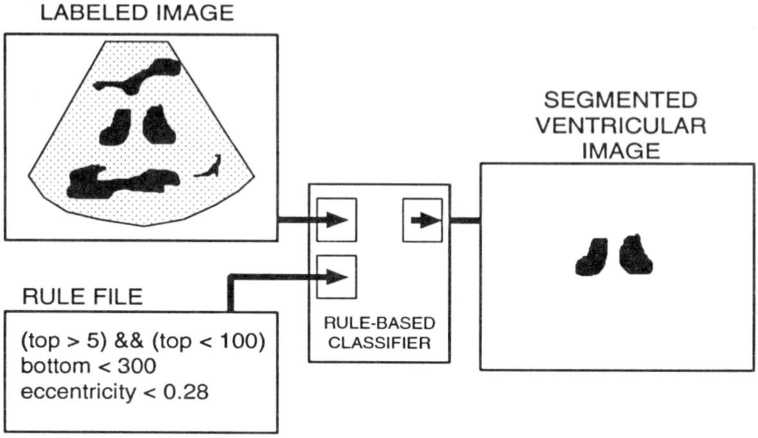

Figure 15: This figure illustrates the use of the vuclass rule based classifier. The classifier takes two input parameters : a labeled image and a rule file. The labeled image contains vectors of information describing the shape characteristics of each region. Regions are rejected if their shape characteristics do not conform to the input rules. The resulting image contains only the desired ventricular regions.

The code that had to be written for the application was greatly simplified by using the Khoros services. Ultrasound images were stored and managed as abstract data objects by data services. The image graphics object available in visualization services provided all the functionality necessary to fulfill the application's image display requirements. Interactive editing was made possible with a raster drawing annotation available with the visualization services. Text graphics objects were used with a large font to create a visually clear status window for reporting information to the physician. Finally, as mentioned earlier, the functionality of the automatic segmentation procedure was implemented simply by making the appropriate series of library calls to the correct operators in the image processing toolbox library. This application was named ultra, see Figure 16.

7.6. Case Study Conclusion

The quantitative volumetric measurement of a premature infant's brain ventricles is possible using information from a radial sequence of ultrasound images. In order to perform this measurement in a reasonable amount of time, a software application was developed. This application utilizes an automatic segmentation procedure for identifying the ventricular regions on each ultrasound image.

This procedure was developed at the visual programming level, using existing Khoros operators to identify a number of regions, of which the ventricular regions are a subset. A rule-based classifier was developed at the toolbox programming level in order

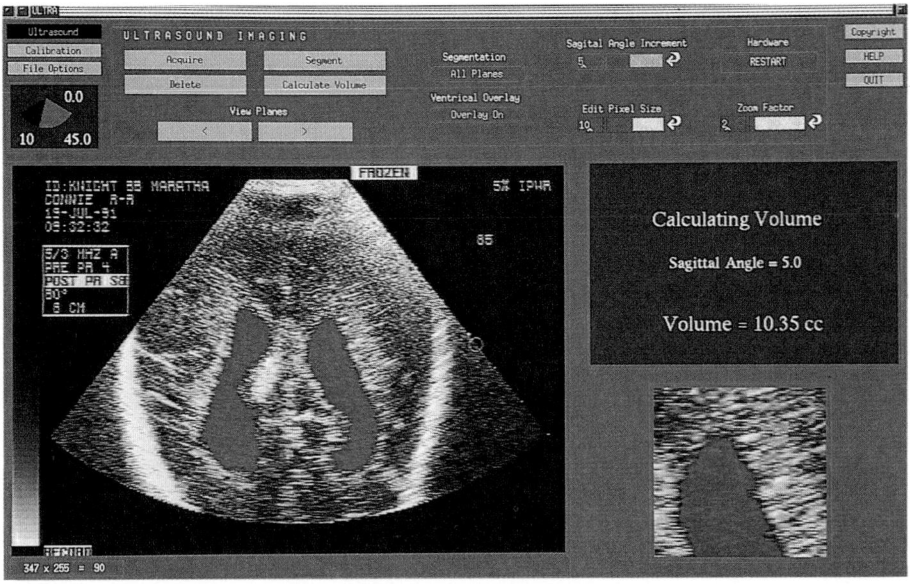

Figure 16: The interface for ultra. At the bottom left is the main image display where each ultrasound image is shown overlaid with its segmented results. At the bottom right is a zoom display which allows for precise checking and editing of the segmented overlay. The buttons *Acquire* and *Delete* allow the physician to interactively maintain a sequence of ultrasound images in computer memory. Once a sequence of images has been acquired, the physician can click on the *Segment* button. The physician can then review the results of the automatic segmentation for correctness, and make appropriate changes. The physician can draw in any regions that the automatic segmentation procedure may have missed, or can erase any extraneous areas which the automatic segmentation may have found. After making whatever minor corrections are necessary, the physician can click on the *Calculate Volume* button.

to classify these regions into ventricular and non-ventricular regions. The final application was also developed at the toolbox programming level. This application is currently being used at the University of New Mexico Hospital's Department of Pediatrics as a functional diagnostic tool.

8. Status, Support, Documentation, Release Dates, Availability

The Khoros system has been applied to a diverse spectrum of applications including process control, medical imaging, automatic target recognition, sensor simulation, ecological research, remote sensing, virtual reality, geographic information systems, telecommunications, and scientific visualization. Khoros continues to evolve as the project introduces innovations in the areas of information processing and data management, software development, distributed computing, graphical user interface technology, and visual programming.

Khoros is being used as a educational tool around the world. It provides a framework for instruction in areas such as geometry visualization, image processing and signal processing. Khoros is accepted as a preferred environment by a large (over 25) consortium of government agencies and industry. The next major increment in acceptance of the system will come with the third-party development of commercial products based on Khoros.

Much of the success of the project can be attributed to the **Free Access Distribution** of Khoros. The Khoros developers involve the community of users in the software development process as much as possible. Release 1.0 of Khoros was offered free of charge via public ftp access on May 10, 1991; the last update was provided March 20, 1992. The public release of Khoros 2.0 is expected to be available in early 1994.

The documentation for the system is a combination of on-line help and printed manuals. The manual is made up of six separate volumes: "Readme First", "Getting Started Manual", "Visual Programming Manual", "Toolbox Programming Manual", "Programming Services Manual", and "Backstage Issues". Also, journal playback files are provided to give the new user "live" demonstrations of the various applications.

For current information on what platforms and architectures Khoros runs on, and for information on how to use anonymous ftp to obtain Khoros, please send e-mail to khoros-request@chama.eece.unm.edu. There is also a USENET news group, comp.soft-sys.khoros, that actively discusses Khoros related issues.

9. Acknowledgments

The research and development that is discussed in this paper has been implemented by the Khoros Group.

Khoros Group, left to right, front row: Danielle Argiro, John Rasure, Donna Koechner, Steven Kubica, Becky Bishop; middle row: Mark Young, Neil Bowers, Tom Sauer, Steve Jorgensen, John Salas; back row: Shawn Larson, Scott Wilson, Ramiro Jordan, Jeremy Worley.

10. References

1. K. Konstantinides and J. Rasure, "The Khoros Software Development Environment for Image and Signal Processing," *IEEE Journal of Image Processing*, to appear in 1993.

2. J. Rasure and C. Williams, "An Integrated Visual Language and Software Development Environment", *Journal of Visual Languages and Computing*, Vol 2, pp 217-246, 1991.

3. J. Rasure, D. Argiro, S. Hallett, R. Neher, M. Teran, M. Young, and S. Wilson, "XVision: A Comprehensive Software System For Image Processing Research, Education and Applications," *ACM SIGGRAPH User Interface Software Symposium*, Banff, Alberta, Canada, October, 1988.

4. J. Caldwell, L Goldberg, and H. Lord, "Overview f AVS5", *Proceedings of the 2nd International AVS User Group Conference*, AVS '93, Orlando, Florida, May 24-26, 1993.

5. Kubica, Brann, Rasure, Sauer, "Procedure for Automatic Ultrasound Image Segmentation of Preterm Infant Brain Ventricles", *Proceedings of the 2nd International AVS User Group Conference*, AVS '93, Orlando, Florida, May 24-26, 1993.

6. Benjamin S. Brann IV, MD, C. Wofsy, PhD, J. Wicks, MD, J. Brayer PhD, "Quantification of Neonatal Cerebral Ventricular Volume by Real-time Ultrasonography

- Derivation and In Vivo Confirmation of a Mathematical Model", *Journal of Ultrasound Medicine, Vol. 9, pp 1-8, 1990.*

7. Anil K. Jain, Fundamentals of Digital Image Processing, Thomas Kailath ed, Prentice Hall, 1989.

8. Scott R. Wilson, "A Vector Quantization Approach to the Restoration of Space Varying Blurs With Application to Ion Beam Optical Figuring", Dissertation, Department of Electrical and Computer Engineering, University of New Mexico, Albuquerque, NM, 1990.

9. Rafael C. Gonzales and Paul Wintz, Digital Image Processing, Addison-Wesley, Reading, Mass., 1987.

Prototyping, Visualization and Simulation Using the Application Visualization System

Carl-Johan Westelius

Johan Wiklund

Carl-Fredrik Westin

Dept. of Electrical Engineering, Computer Vision Laboratory, Linköping University
S–581 83 Linköping, Sweden
email: cawe@isy.liu.se, jowi@isy.liu.se, westin@isy.liu.se

ABSTRACT

The Application Visualization System software from Advanced Visual Systems Inc is an interactive visualization environment for scientists, engineers and technical professionals. This report contains a short overview of the AVS software packages and a discussion about its general performance. The software package has actively been used at the Computer Vision Laboratory, Linköping University, during the last three years. The AVS package has been used in many applications. Examples are generating images from a virtual environment, simulation of a controllable robot with a stereo camera head and visualization of multidimensional data structures. Lately we also have used AVS for handling communication between different processes which may be distributed on different machines. AVS was primarily developed as a tool for visualization of complex data sets. However, another important aspect of the software is that it can be used as an advanced workbench for controlling networks of Unix processes (including external ones on different machine types) using simple visual programming.

1. Introduction

This article is intended to communicate the experience gained at the Computer Vision Laboratory using the AVS software for algorithm prototyping, data visualization and graphic simulation. The first section contains a short background and perspective summary of visualization techniques in general. Then follows an introduction to the different parts and data structures in the Application Visualization System, AVS. AVS module communication and execution is thoroughly described in section 4. The second half of this article describe how AVS is used at the Computer Vision Laboratory, the benefits and pitfalls, and concludes with suggestions of improvements.

2. Background on Visualization

Visualization has become an important component of the scientific discovery process, allowing engineers and scientists to examine large amounts of data with higher efficiency and greater comprehension[6, 1]. The problem is to communicate the information content in the data to scientists so that they can use it effectively in the human creative and analytic processes. This requires a method of communication with a high bandwidth and an effective interface. Scientific visualization uses the human vision system and computer generated images. With color, intensity, transparency, texture, and a number of other techniques, images may bring a tremendous amount of information in a short period of time.

A less mature issue is what to do when the problem's dimensionality exceeds what we can handle using traditional methods. There is a need to develop mathematical methods to analyze and visualize multivariate data [8]. There is both need and interest to extend many of the univariate, one-dimensional methods of data fitting and data reduction to higher dimensions and more general domains. New visual techniques are emerging to extract the nature of higher dimensional data sets or functions. One example, where research has only begun, is the problem of interpolating scattered data over arbitrary surfaces. Still needed are robust, reliable methods to handle very large scattered data sets of high dimensions. Another example is volume rendering. It is very computation intensive, and many applications require interaction. Real-time volume rendering is not possible presently, but due to the potential benefits, it will surely to be a reality in the near future.

Computational fluid dynamics (CFD) is an important scientific topic because of its widespread applications. CFD has emerged as a key visualization area precisely because computer power now permits the large simulations required to model complex processes. CFD results are often complicated even for a single time-step and tend to be very dynamic. Animation has shown to be a useful tool for the understanding of these complex data. The techniques used to track features of interest in a CFD simulation illustrate the ongoing fusion between computer graphics and digital imaging.

New methodologies to guide the selection of visual representation of data are needed. Understanding how cognitive perception operates will help defining the limitations of visual methods. While no single approach is likely to apply to all visualization problems, developing workable paradigms will help us move towards design tools for displaying data. Eventually, such paradigms can lead to a sophisticated automation capability. Current research into the use of sound, eye-motion, and interactive hardware devices attempts to define what can be achieved in a multisensory environment. Exciting research is exploring improved user interaction with the data analysis process by using virtual environments. In a virtual environment the user, through a variety of input sensors, enters the scene and roams the data. Upcoming research will explore the benefits of such interaction. To meet the needs of the visualization community, a new conference series was initiated by the IEEE Technical Committee on Computer Graphics in cooperation with ACM Siggraph. The first conference, Visualization 90 [4], took place in 1990 in San Francisco.

There are several software packages/environments for the general display of scientific data. These include apE by the Ohio Supercomputer Center [2]; Khoros developed at the university of New Mexico [3]; AVS (Advanced Visualization System) by Advanced Visual Systems Inc. [9, 3]; Explorer by Silicon Graphics [3]; PV-Wave by Precision Visuals [7]; VDE (Visualization Data Explorer) by IBM; Sun Vision by Sun Microsystems; Voxel View by Vital Images; RIP (Real-Time, Interactive Particle Tracing), GAS (Graphical Animation System), and Plot 3D developed at NASA Ames; and Data Scope, Image Tool, and X Data Slice from the National Center for Supercomputing Applications. Of particular interest are the dataflow concepts and visual programming paradigms shared by apE, AVS, Explorer and Khoros. The basic idea here is that the user can interactively create a dataflow network by selecting and connecting icons representing various modules.

Interactive input devices are continually improving. Common 2D devices include knobs, switches, pedals, mice, and tablets. Tablets are the most general device and also need most improvement; they need higher resolution, higher speed, and more degrees of freedom. Six-dimensional interactive devices are also available, providing the usual 3D positional information plus three degrees of orientation information (yaw, pitch, and roll). Higher-dimensional devices, such as the data glove, have begun to appear. These will improve to offer higher resolution, higher speed, and lower cost.

Raster displays of 2D frame buffers have improved steadily to offer more colors, higher resolution, and less flicker. High-definition television (HDTV) will affect visualization. Also, video is moving towards an all-digital format, designated 4:2:2, to standardize digital interconnections of diverse video products. HDTV will urge the development of compatible recorders. Film recorders will become cheaper as the technology becomes cheaper and the competition matures. Should stereo become a widely accepted mode of presentation for volume visualizations, then stereo film and video standards will have to be developed. Graphical workstations now have full hardware support of surface graphics, offering rendering features such as texture mapping, bump mapping, antialiasing, reflections, transparency, and shadows. Among image processors, fast planar machines have existed for some time. These machines contain special boards for certain aspects of image processing. Faster versions are becoming available that have wider processing capabilities and higher resolutions. In fact, the notion of a general-purpose image processor that can implement any image processing algorithm as a program is becoming common.

3. General Description of the AVS Software

The AVS software is a product from Advanced Visual Systems Inc. (AVS Inc.). It is an interactive visualization environment for scientists, engineers and technical professionals. AVS offers an environment that allows users to construct applications incorporating their own code without graphics programming. The package is easy to learn and use, it provides an intuitive interface for quickly designing prototype applications and it supplies powerful tools for customization and tuning production applications.

There are two major ways of using AVS. Firstly, AVS has a number of data visualizers called viewers. These are ready-to-use visualization packages for a variety of data types.

- *Image viewer* is an interactive 2D image processing and display package.
- *Geometry viewer* is an interactive 3D geometric data renderer.
- *Graph viewer* is a tool for plotting functions, measurements, statistics using line, bar, area, scatter and contour plots.

Secondly, AVS can be used as a prototyping tool. New algorithms and methods can easily be designed and tested using visual programming, i.e. choosing and interconnecting program modules graphically, see figure 1. The prototyping subsystems are:

- *Network editor* is a visual programming tool.
- *Module generator* is an integration tool for user-supplied code.
- *Layout editor* is a graphical user interface design tool.
- *Command Language Interpreter* is a scripting language and callable interface.

This report is concentrated on the prototyping capabilities of AVS.

AVS users can construct their own visualization applications, by combining software components into executable flow networks. The components, called modules, implement specific functions in the cycle. The flow networks are built from a menu of modules by using direct-manipulation, a visual programming interface called the AVS Network Editor. With the Network Editor, the user produces an application by selecting a group of modules and drawing connections between them.

The AVS Command Language Interpreter (CLI) is a text language that can drive most of the AVS system. It can be used for making animations, save networks, remote control, etc. Any module can pass CLI commands to the AVS kernel, i.e. any module can modify parameters, rearrange networks, invoke and delete other modules, etc.

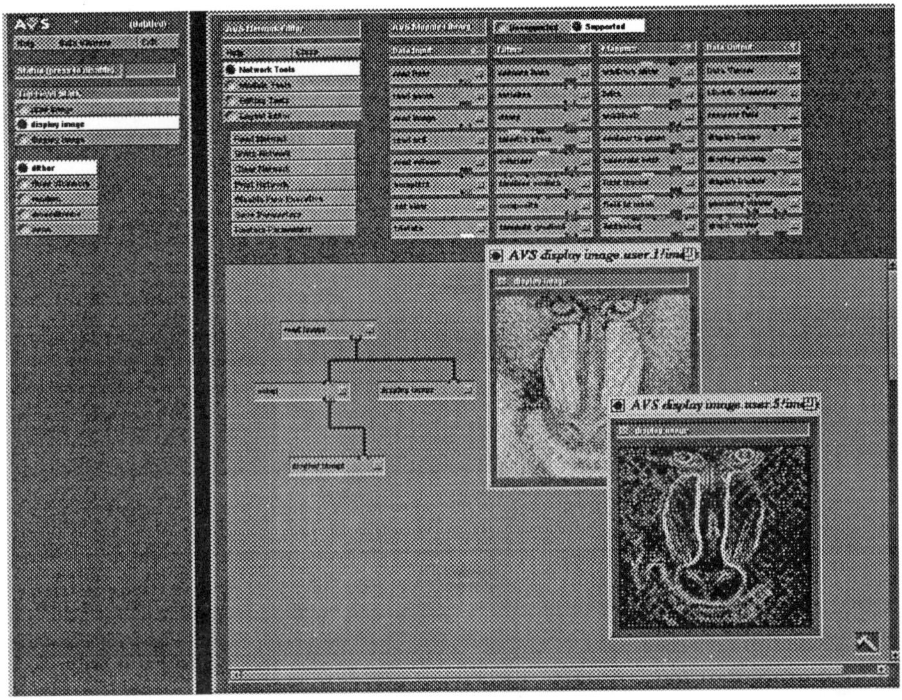

Figure 1: The AVS network editor. The window to the left is the network control panel. This is the default location for module widgets. The large window is the network construction window. The upper part contains a network editor menu to the left, and a palette of modules in the current library to the right. The rest of the network construction window is a workspace. The sample network reads an image from a file, displays it, runs a sobel operator and displays the result.

AVS includes a rich set of modules for the construction of networks. It allows users to create their own new modules to meet their specific needs and dynamically load them into networks. A special module generator can be used to automatically generate module code in FORTRAN or C. Both ANSI–C prototypes and C++ classes are available. The module generator also creates Makefiles and manual pages.

The basic data types in the system are oriented towards scientific data manipulation and graphic display. The types include:

Field data is an n-dimensional array with an m-dimensional vector of values at each array location. The vectors can be of bytes, shorts, integers or floating-point values. The grids can be uniform, non–regular and rectilinear. It is also possible for the grid to describe a curved or arbitrarily deformed space (curvilinear), or an arbitrary list of point in 3D space (scatter data).

Unstructured cell data (UCD) consists of a geometric model built of individual cells. The cells, including 1D, 2D and 3D shapes, are defined by nodes, or vertices. Cells may be points, lines, quadrilaterals, triangles, tetrahedrons, pyramids, prisms or hexadrons. Data, either scalar or vector, can be assigned to the entire model, to each individual cell and to each vertex of each cell.

Geometric data consists of three-dimensional objects that are constructed out of one or more of the following primitives: polyhedrals, polygons, meshes, spheres and polytriangles. AVS passes geometric information between modules by using a data structure called an *edit list.* The edit list describes changes to the geometry of a scene. Generally a user module sends edit lists as outputs. Geometry output is typically used as input to an AVS-supplied renderer module such as the geometry viewer.

Pixel Maps are data structures that incorporate references to X Window System pixmaps. An X pixmap is an array of pixel values that can be destination for a rendered image. It resides in the X server. (In contrast, an image is a data structure that includes an array of colors and resides in client memory.) Because pixel maps rely heavily on specific hardware and software features, they are not very portable or easy to use. Programmers should not try to use pixel maps to perform image processing; the "image" field type ("field 2D 4-vector byte uniform") is more portable and interfaces to a wider variety of other modules.

Colormap is an arbitrary-sized one-dimensional array of 4-dimensional vectors. Vector components include normalized floating-point values for hue, saturation, brightness, and opacity.

User defined data. AVS provides limited capabilities for users to implement their own data types. User-defined data types may be useful for problems that are best defined using data structures that are different from those built into AVS. However, existing modules are unlikely to be able to deal directly with the new data type; the user has to convert to a more standard type eventually. The user-defined data type may also be useful for sending a subset of data back "upstream" in a network to feed back information to a module that is sending data "downstream".

Molecular data type (MDT) consists of ten defined CHEM objects that can be combined hierarchically in linked lists to represent the structure of molecules.

3.1. Hardware, Operative System, and Language Support

AVS currently operates on all the major UNIX workstations. For massivly parallel algorthms it is also possible to run modules remotely on Meiko, Thinking machines and Wavetracer.

AVS has support for remote X-window displays, AVSX, which means that low-cost, 8-bit X-terminals and workstations can be used.

Programming can be done in K&R C, ANSI–C, C++ and FORTRAN.

3.2. Module Libraries

About 150 standard supported modules are included in the AVS software provided by AVS Inc. In addition to this, a number share ware modules are available. The International AVS Center, housed at the North Carolina Supercomputing Center, is a center for the collection and distribution of new modules. Hundreds of user-contributed and ported public domain modules are available and can be retrieved via ftp (including documentation). Examples of recently ported public domain modules are *Khoros* (approx. 250 modules) and *Sunvision.* A database facilitates searching for modules suited for certain applications. The center also distributes a magazine quarterly, the AVS Network News magazine, presenting novelties concerning AVS software and applications.

Our own CVL library contains about 100 modules developed specially for image processing problems. The robot simulator described below is a part of this library, c.f. section 8.

4. AVS Module Execution and Communication

AVS modules can run locally on the same host as AVS or remotely on other hosts, possibly with a different architectures. Separate modules may also run in the same unix process saving resources and enhancing overall performance.

AVS5 uses direct module communication which optimizes network-based data transfer by passing data directly between modules using peer-to-peer socket connections. The most efficient communication mechanism to use for data transfers between modules is automatically selected based on the following priority:

1. Pass a memory pointer for modules in the same unix process.
2. Pass a shared memory ID for modules on the same system but in different unix processes.
3. Use a direct socket connection between modules on different systems and systems without shared memory.
4. Use the XDR data format and direct socket connection between modules on different systems with incompatible data formats.

Two communicating remote systems will pass data directly using sockets. Control and data are separated allowing for optimization of each. The socket to the AVS kernel is used for transferring control information, while data is transferred directly between the modules without a middleman. AVS handles all details of the underlying communication mechanisms used. The communication mechanism to use is dynamically determined on a per-communication basis. All AVS systems can share data with necessary conversions performed transparently to the user. Communicating systems with incompatible data formats will convert to XDR data representation.

4.1. Interface to External Processes and Dedicated Hardware

There are many examples of having AVS modules controlling external hardware. At the Computer Vision Laboratory in Linköping, modules for controlling data flow to a video disk (Abekas) has been developed. A module interfacing towards Motoman robot arm and a framegrabber has also been developed. Actually every external process that can be communicated via sockets can be driven from a module in AVS.

It is also possible to control AVS from external processes. When AVS is started in server mode it displays a message, giving processes and port number. Command Language Interpreter commands can then be supplied through the port from any other unix process, see section 3.

4.2. Module Execution and Synchronization

The execution of modules in an AVS network is controlled by the netflow executive. It is a data flow based control structure, i.e. downstream modules execute when triggered by output from upstream modules. Modules can also be executed asynchronously with the rest of the network. If the network has branches, the different branches can execute in parallel. With this control strategy, single stream modules are executed serially.

4.3. Parallel Multihost and Multiprocessor Execution

Parallel module execution enables independent modules to execute concurrently. Multi-processor systems can execute module groups concurrently on separate CPU's. Remote modules can be assigned to separate systems utilizing the processing power of the entire computer network. AVS's use of processing resources is controlled by specifying the maximum number of concurrent processes.

Figure 2 shows a network with three independent data streams. The data from "*read image*" is split into three "*ip convolver*" modules, each with a different filter supplied by the "*ip read kernel*" modules. These three data streams can be processed on three different machines or CPUs. Suppose the "*ip arithmetic*" module and subsequent modules are running on the same host as the the leftmost data stream. The "*ip arithmetic*" module will then automatically wait for the middle stream to finish before it begins executing.

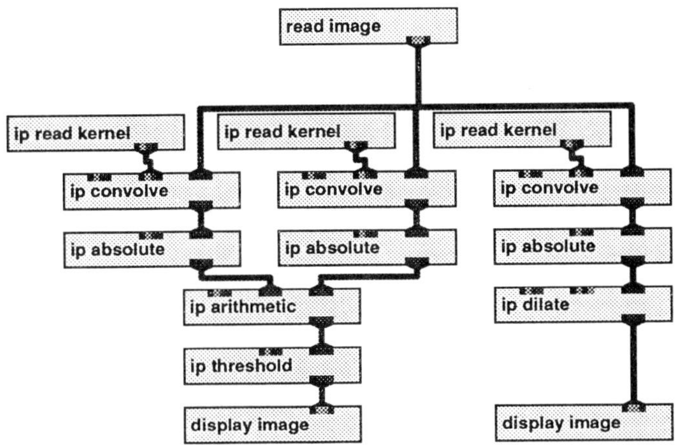

Figure 2: A net suitable for parallel multihost or multiprocessor execution.

4.4. Selective Computation

When a module has more than one input port or parameter, it is possible that, when the module computation function executes, some ports or parameters have not changed since the previous execution of the computation function. Sometimes it is also necessary to wait for all input ports to change in order to get synchronized inputs.

AVS provides routines to determine what has and what has not changed, which enables a module to avoid computations on ports or parameters that have not changed. AVS also provides a routine to declare that a given output port has not changed since the previous invocation of the computation function, which reduces downstream module execution. These two functions are very important when creating loops in the module network. The network in figure 3 is used to probe volumes of data. The "*probe*" module generates a probe cursor geometry, but it needs information about the position of the probe in order to supply the corresponding value from the data field. This is handled with the upstream connection from the "*geometry viewer*". A loop is created and special considerations have to be taken in order to avoid an infinite loop locking the whole net. Going through the net step by step explains how the infinite loop is avoided:

1. "*read field*" reads the data field from a file, and sends it to "*isosurface*","*volume bounds*" and "*probe*". These three modules generate geometries corresponding to the isosurface, the volume boundaries and a probing device, and sends them to the "*geometry viewer*", see figure 3.

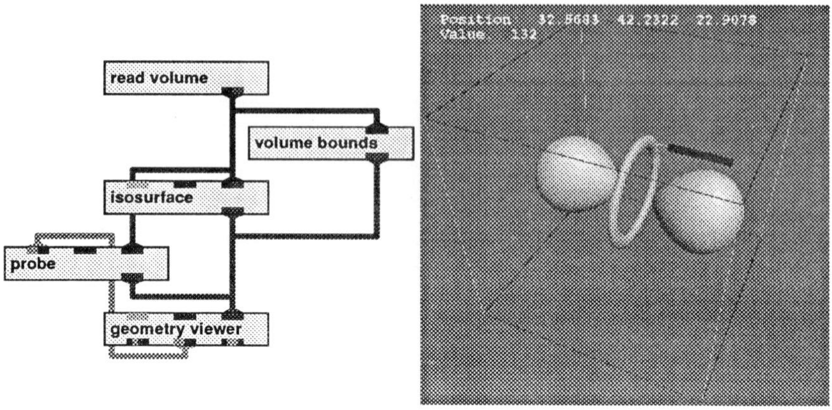

Figure 3: Data field probing network and the "*geometry viewer*" output image. The upstream connection from the "*geometry viewer*" to the "*probe*" is made automatically and is normally invisible.

2. The "*geometry viewer*" renders the image from the geometries and marks its upstream output port as unchanged. The system is now "idle".

3. If the user selects and moves the probe geometry, the transformation is sent upstream. The "*probe*" module now gets a new value from the data field and supplies new geometries to the "*geometry viewer*".

4. The "*geometry viewer*" renders the image from the new geometries and marks it upstream output port as unchanged. The system is now "idle" again.

If the "*geometry viewer*" had not marked the output port as unchanged when receiving geometries, the "*probe*" would have executed once more, triggering the "*geometry viewer*" again and so on. Thus, the two modules would trig each other infinitely.

4.5. Subroutines and Coroutines

There are two different types of AVS modules; subroutines and coroutines. Subroutines are the most common type, they compute only when input data or a parameter has changed. Coroutine modules usually perform independent computations, and can interact with AVS whenever wanted. Coroutine modules can be set to execute synchronously, i.e. under the control of the flow-executive, or to execute asynchronously, i.e. they can run independently of other modules.

Coroutine modules have the ability to execute asynchronously from the AVS flow executive. This means that at any time coroutine modules can acquire their input data and send output data. This provides coroutine modules with more flexibility than subroutine modules. For example, they can manage their own input sources (such as X events or keyboard input), they can run simultaneously with other AVS modules, and they can schedule execution themselves rather than waiting for user interaction.

4.6. Asynchronous Execution: A Process Control Example

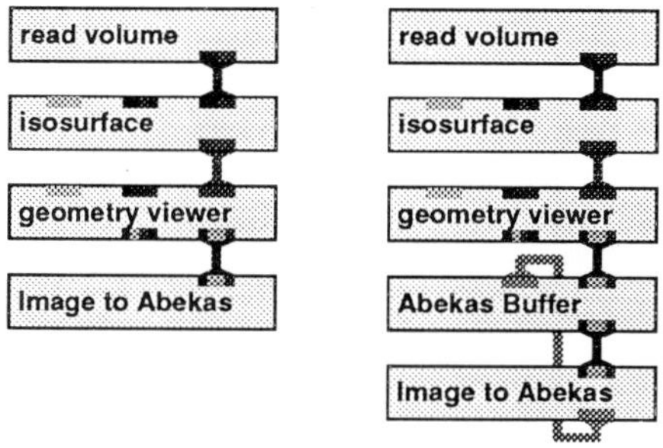

Figure 4: Left: A simple net of subroutines executed serially. Right: "*Abekas Buffer*" will hold the net until the upstream signal indicates that the asynchronously running "*Image to Abekas*" is ready to receive the next image.

Consider the following example of how different module types can be combined to reduce execution time. The left hand network in figure 4 is a number of modules, serially executed by the flow executive. The last module, "*Image to Abekas*", rearranges the image data input (ARGB) to a format suitable for the abekas videodisc, and does the actual data transfer to the abekas via unix socket calls. The total execution time for the "*Image to Abekas*" module varies between 5 and 30 seconds depending on the type of data sent. For BW (1 byte) pixel data it takes about 5 seconds, and for RGB (3 byte) pixel data it takes about 30 seconds. For each loop in this network, the rest of the net must wait until the data transfer to the Abekas is finished. This is a typical case where it would be more efficient to run modules simultaneously. If the rest of the network could work on producing the next image while the current image is transferred, up to 30 seconds per frame could be saved depending on the relation between image generation time and image transfer time. When generating a 30 second video sequence (750 frames), up to 5 hours could be saved.

In order to make the "*Image to Abekas*" module in the example above to run simultaneously with the rest of the net, it must be implemented as a coroutine module that executes asynchronously. This gives raise to a synchronization problem. We do not want to loose images produced by the first part of the net or send the same image to the Abekas multiple times. By introducing a new module, "*Abekas buffer*", that manages the synchronization between the flow-executive and the asynchronously running "*Image to Abekas*" module, this problem is solved. See the right hand network in figure 4.

The module "*Abekas buffer*" is implemented as a synchronized coroutine. It releases the AVS net as soon as it has received an image, then it waits until the "*Image to Abekas*" module signals that it is ready for the next image. The ready signal is sent via the upstream connection from the "*Image to Abekas*" to the "*Abekas buffer*" module in figure 4. This connection is normally invisible and connects automatically when the image output-input ports between the two modules are connected. When the ready signal arrives, the image is put on the output port. If a new image appears on the input port before the ready signal has arrived, the "*Abekas buffer*" module holds the flow-executive until the " *Image to Abekas*" module is ready. There is no AVS function call available that holds the flow-executive. We have implemented it as a loop in which the input ports are polled. It would be better to wait (while holding the net) for a signal telling that the image transfer to the Abekas is ready.

4.7. Pipelining

The normal flow based execution control combined with the capabilities of selective execution are enough for most applications. There are, however, ways to make modules execute in any order.

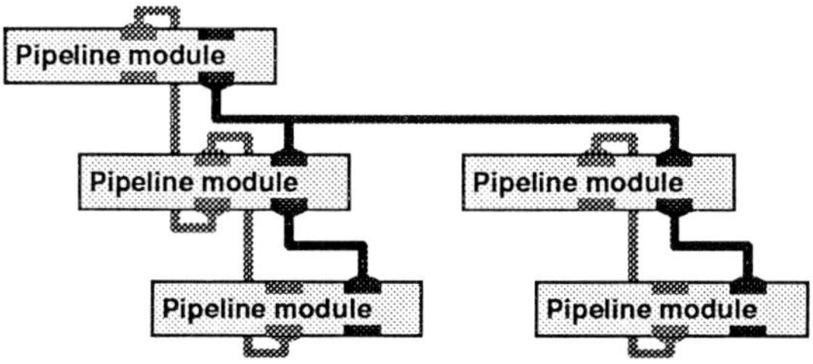

Figure 5: An AVS pipeline network example.

In the subsections above it was described how the AVS flow executive can handle execution scheduling of networks with parallel streams of data, unsynchronized modules and data flow loops. There is, however, no support for piplining modules, i.e. having modules in the same data stream working simultaneously on data from different "time instants".

A pipeline structure can be constructed with asynchronous modules and an extension of the "ready–signal" system described in subsection 4.6.. First, consider the left data path in figure 5. "*Pipeline module*" is a dummy module representing any asynchronous module both taking a ready–signal from the downstream module and giving a ready–signal to the upstream module. The signal system allows modules to hold upstream modules if they are not ready to receive data. If the modules execute on different processors, it is possible to get the whole network to execute at the rate of the slowest module plus possible communication overhead.

The approach chosen above has two major problems. Firstly, it is not possible to use standard modules in such a net. Secondly, it is not possible to branch the dataflow into an arbitrary number of branches. In figure 5 the topmost module in the right branch has nowhere to connect its ready–signal, since the input port of the upstream module is already occupied. This could be handled by letting input ports have multiple connections, but there is no support for that at present. There are, however, standard modules capable of taking multiple connections to the same input port, so it may be supported for user modules in a future release of AVS.

4.8. User Defined Execution Schedule

Since any module can control almost any aspect of AVS, a special controller module can be constructed. The controller module can invoke or delete modules, change module interconnections and enable or suppress module execution. Thus, it is possible to build a net by just invoking one controller module, which then invokes sub-controller modules, possibly on other hosts, which in turn invoke the actual processing modules. The main controller can therefore control processes without knowing all the details about the modules included in the subnets.

5. AVS Usage at the Computer Vision Laboratory

AVS, version 2, was introduced at the Computer Vision Laboratory in 1991 in connection with a purchase of a Stardent GS2500 computer. Initially, the staff was reluctant to use AVS2. One reason for this was that the input/output of the local file format needed special modules, and at that time the "*Module Generator*" did not exist and the documentation was not as good as today. Another important reason was that while the supply of 3D data visualization was satisfactory, the lack of interactive 2D data visualization, compared to the existing GOP–system, made people hesitate about the benefits of changing systems.

The GOP–system consists of special hardware to perform complex arithmetic convolutions, and special hardware to display complex images. The convolution hardware consists of two filter processing pipelines and a floating point pipeline. It is connected to a SUN host computer via a Sbus–VME bus connector. There is a vast amount of software support to configure and use the special hardware, for instance image operations are defined by pascal-like programming. The functions in the display unit are controlled via a menu interface. The GOP–system was originally built to handle 2D data, but now it exists software for doing convolutions of up to 4D data. The display unit, however, still operates on 2D images.

Today, a number of AVS modules, for I/O and 2D visualization, have been developed which, together with new AVS5 standard modules, furnish an environment compatible with the visualization capabilities of the GOP–system for 2D data and which is much better for 3D data, c.f. section 7.

AVS is used in a number of different projects in basically three different ways at the Computer Vision Laboratory:

1. Algorithm prototyping.

2. Data visualization.

3. Graphic simulation.

These three modes of application are often combined, e.g. data from a prototype algorithm is visualized using the standard tools of AVS, or a graphic simulation feeds a prototype algorithm with test data.

6. Algorithm Prototyping

With algorithm prototyping is meant both the development of basic operations, such as sobel filtering or local orientation estimation, and the combination of such operations into more complex algorithms.

In the GOP-system both the set up of the convolution hardware and the floating point operations are defined in a PASCAL-like language. The combination of operations into complex algorithms is handled with shell-scripts and operations are communicating via files. This makes the GOP–system suitable only for still images, volumes and pre-recorded image sequences. Furthermore, the code is not portable to any other architecture.

In AVS the operations can be written in K&R C, ANSI–C, C++ or Fortran. Although the AVS specific parts of the code are not portable to anyone without AVS, the rest of the code can be written as library functions enabling porting to other systems.

6.1. Creating and Debugging AVS Modules

The "*Module Generator*" is a special module for generation of new modules. Input ports, output ports and parameters are defined using point and clicking. In the present version all ports and parameters have to be defined from scratch. When having a number of ports or parameters of similar types a copy command would save much of tedious mouse clicking. The "*Module Generator*" provides a source file with marked areas where to include your own code. Host–independent Makefiles are also created. Unfortunately, there is no easy way of adding libraries or object files when compiling this way. The Makefile is rewritten before compiling and any changes made by the user are lost. A manual page skeleton can be created automatically having the information about ports and parameters pretyped at appropriate places. The manual page may then be included in the AVS on-line help system either manually, by editing the ".topics" file, or automatically by running AVS with the "-reindex" option.

The "*Module Generator*" is a fairly new feature in the AVS system and there are still things to be improved. A serious problem is that it sometimes produces incorrect code. The bugs are easy to spot for an experienced user but for a beginner it might be harder. Another problem is that lines of code longer than 70 characters are destroyed by the "*module generator*" if the code is re–read for editing. These problems are reported as known bugs by AVS Inc. and hopefully they will be gone in the next release.

Due to these malfunctions the "*Module Generator*" has only been used to generate skeletons for the code, makefile and manual pages. All developing and debugging has been carried out not using it.

AVS provides a facility, avs_dbx, for debugging a module during the execution of an AVS network. A shell script file arranges for a module to run under the native debugger. On many UNIX systems, the native debugger is *dbx*. It is also possible to specify an alternate debugger, e.g. gdb. The avs_dbx is mainly a positive experience. The starting up procedure might seem a bit complicated to the beginner but the cookbook example in the documentation sorts things out.

Another important tool is memory allocation debugging. This tool consists of a layer of functions between the application and the system memory allocation routines. By using these functions, the following problem can be tracked down:

- Write beyond end of allocated memory.
- Use of memory after it has been freed.
- Assume that *malloc* clears memory.
- Not checking for NULL returned from *malloc*.
- Specify single argument to *calloc*.
- Failure to free memory ("a memory leak").

All source files using malloc, calloc, etc, have to include at least one of the header files supplied by AVS. This might cause problems when having function libraries. The memory allocation debugging function has only been tried a few times a the CVL. The important information is drowned in the information flow about all internal memory allocations of the AVS system. There is an environment variable controlling the amount of information to be extracted but it did not seem to work.

AVS5 generally sends good error messages when module execution in a network fails. This is certainly very useful in the debugging procedure.

6.2. AVS Networks and CLI Scripts

The AVS modules are connected into networks which can be saved as network files. The module parameters and widget layout are also saved. Viewer parameters, on the other hand, are *not* saved in a network file. This means, for instance, that camera parameters can not be restored when saving a network including the "*geometry viewer*".

An alternative way of saving a network is to use the Command Line Interpereter, CLI. Almost all of AVS can be controlled using keyboard commands. AVS can record user actions as CLI-commands in a ASCII script file. The script file may then be edited using an ordinary wordprocessor or editor. Unfortunately, the viewer commands can not be recorded, but they may be added manually when editing the script file. For more complicated networks a CLI script file is to prefer. In addition to the possibility to set viewer parameters, module parameters can be set in any order ensuring the proper set up sequence. Network files set all the parameters and executes the modules from the top and downwards. Especially in networks with feedback connections this is not always the wanted sequence.

7. Data Visualization

The GOP–system has a dedicated display processor for fast and easy display and interrogation of 2D scalar and complex data (in polar representation). The GOP features include:

Real time zoom and color table manipulation. Re-mapping and color slicing in real time give an intuitive understanding for the qualitative features in an image or data field. These operations can be applied to either the magnitude (controlling the intensity) or the argument part (controlling the color) or a combination between them. Having got used to such facilities it is hard to do without them.

Data statistics. Interactive, mouse controlled histograms, 1D profiles and pointwise probing give information about quantitative entities in the data field.

Image modification. It is possible to interactively paint in the image with different colors and brushes. This function can also be used to mark a region of interest (mask) for following display processor functions.

Image calculator. The calculator parses a user defined expression including image files, functions and constants. The expression is evaluated in the GOP floating point processor.

7.1. Real Time Zoom and Color Table Manipulation

The GOP–system relies on a special purpose display processor while AVS uses the X-windows system. The display is thus as fast as the X-server used. Modern workstations are closing in on the GOP–system with regard to the display functions.

7.2. Data Statistics

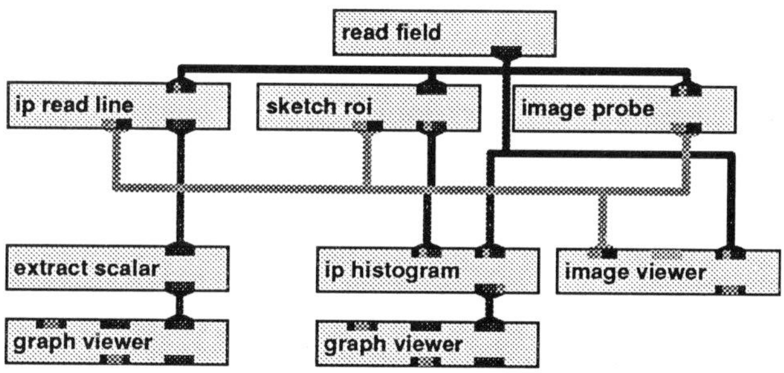

Figure 6: A network for interactive probing of a data field. Invisible, auto-connected upstream connections supply the probing modules with information about the mouse actions in the "*image viewer*" window.

Earlier versions of AVS (*ver.* ≤ 4) did not have any good interactive facilities concerning data field statistics. Researchers at the CVL felt uncomfortable with this and returned to the GOP–system. In version 5 AVS included the modules "*image probe*", "*ip read line*", "*ip histogram*" and "*ip statistics*" which allow the user to get data from one point, from points along a line and, finally, histograms and statistics within an optional region of interest (ROI).

Unfortunately, there are inconsistencies in how the modules handle vector data. The "*ip histogram*" and "*ip statistics*" modules have a choice widget for selection of which vector component to use. The "*ip read line*" module, on the other hand, extracts a 1D vector field requiring an extra "*extract scalar*" module, see figure 6. It would be better to have the same system on all related modules. Skipping the choice widget and altering the "*graph viewer*" to accept vector data as multiple plots, it would be possible to get a simultaneous histogram of, for instance, the red, green and blue content of an image.

7.3. Image Calculator

The GOP–system is equipped with a flexible image calculator enabling the user to perform any arithmetic expression on an arbitrary number of image files by typing an ASCII-string. There are also a set of basic functions, e.g. abs(), sqrt(), exp(), log(). There are also a number of image related functions, e.g. col() and line() which vary between 0 and 1 from the left hand side to the right hand side and from the top to the bottom of the image respectively. For example:

```
outim := rect(line(),0.25,0.75)*sqrt(0.5*(sqr(inim1)+sqr(inim2)))
```

The expression above would create an image file "outim" which is the RMS of the two input images within a centered horizontal band. The band is half the image size in height, since $0.5 = 0.75 - 0.25$. The output image is zero outside the band.

The existing field arithmetics modules in AVS take two identical input fields and perform one mathematical operation on them, e.g. componentwise multiplication. Composing an arbitrary arithmetic expression is both tedious and inflexible. A module that takes any number of not necessarily identical fields and produces any type of field specified by an expression given by the user would reduce the size of many networks. However, this would also require enhanced editing facilities in the text widgets, c.f. section 10.

8. Graphic Simulation

A robot vision simulator has been developed in AVS to facilitate testing of robot vision and control algorithms.

The simulator reduces the need for expensive special purpose hardware since it can be run in "slow motion". A real-time process can be investigated although only limited computational resources are available. It also allows testing of different types of robots and robot configurations without any extra cost. The scene can be varied from a few very simple polyhedral objects to complex, realistic, texture mapped environments. The simulated reality, such as true 3D-structure, true distance etc, can easily be compared to the results obtained by the robot in order to evaluate the performance of the algorithms.

8.1. Texture Mapping Problems

Texture mapping enhances the naturalistic look of simulated objects since the surface structure of real objects can be captured. It allows having detailed structures even though the objects are defined with a few polygons.

Almost all computer graphics systems, AVS included, use linear interpolation when calculating coordinates for textures, which may make a flat surface look warped. The effect can be reduced, when having still images, by making a finer tessellation of the surface, but for images sequences, e.g. from moving cameras, this method is not satisfactory. Textures seem to float around when the camera is moved.

In principle it is possible to add the texture when creating the objects and create one polygon vertex for each pixel in the texture map. The objects will then consist of patches colored according to the texture, as opposed to color the projected image of the objects. This method yields better results but is very memory and time consuming. The texture maps should be done using the accurate hyperbolic interpolation instead. At least, there should be a hyperbolic mapping option.

8.2. Robot Vision Simulator Modules

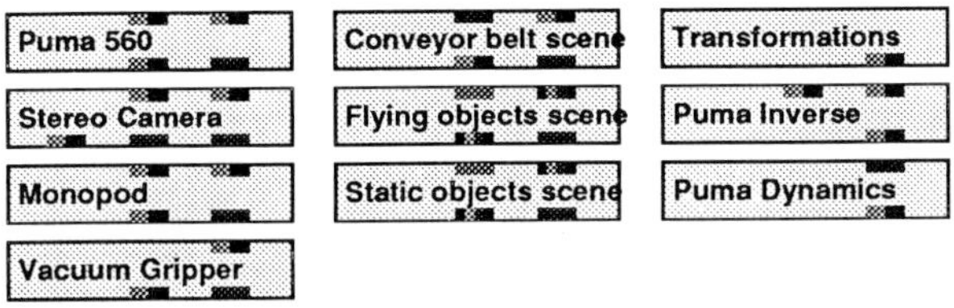

Figure 7: The basic modules used in the Robot Vision Simulator

A set of modules for geometric description and control of robots and tools has been implemented.

"Puma 560" generates a geometric description of a puma robot arm with six degrees of freedom. The robot can be positioned by supplying a transformation matrix to an input port. The robot pose is controlled via the joint angle parameters. There are basically three ways of furnishing joint angles. Firstly, they can be given manually using the widgets on the control panel. Secondly, a controlling module can be attached to the remote control input port, passing joint angles using a user defined data format. Finally, there is also a possibility for a module to access the joint parameter widgets using the "AVScommand" function call.

The module *does not* simulate the dynamics of the arm. In order to implement robot dynamics and/or other forms of control parameters, a filter can be inserted between the controlling module and the "*Puma 560*" module.

The transformation matrix describing the location and orientation of the actuator end point is presented on an output port, enabling attachment of different tools, e.g. the vacuum gripper or the stereo camera head. In fact, it is possible to put one robot arm in the end of another.

"Stereo Camera" creates a geometric description of a binocular camera head. The cameras can be rotated independently and the field of view can be varied. The camera parameters can be manipulated in the same manners as the robot arm.

The camera head is attached to the robot arm by connecting the transformation matrix input of the "*Stereo Camera*" to the corresponding output of the "*Puma 560*". The actuator end point matrix is then transferred to the camera head, making it follow the arm movements.

Note, that "*Stereo Camera*" has two output geometry ports, since two "*geometry viewers*" are needed to get two output images.

"*Monopod*" is a camera head platform with three degrees of freedom: pan, tilt and elevation.

"*Vacuum Gripper*" is intended to be used together with the "*Flying object scene*", "*Static object scene*" or "*Conveyor belt scene*" below. It creates a geometric description of a suction cup to be positioned in the end of a robot arm. As the name indicates, the function is the same as lifting small objects with a vacuum cleaner hose. The module has a boolean parameter indicating whether the gripper is active or not. When activated the transformation matrix is sent to the output port.

"*Conveyor belt scene*" has two conveyor belts, each connected to a machine. One machine produces objects (cylinders and boxes) and puts them on the belt. The selection of type of object and creation time is externally controlled. At the end of this belt there is a bin collecting manufactured objects. Another belt transports objects placed there to a machine where they are consumed.

The "*Vacuum Gripper*" output can be attached to the module enabling manipulation of the manufactured objects. For the robot to lift an object, it has to position the gripper close to the bounding box of the object, orient it perpendicularly to the surface, and activate it.

Objects follow the conveyor belts when put on them and fall to the floor if released in the air. Objects can, however, be moved through each other.

"*Static object scene*" generates a geometric description of a room with two pictures on each wall and light tubes in the ceiling. A table, with a cube, a cone and a cylinder on top of it, is placed in the middle of the room. The objects on the table are pickable and movable with the "*Vacuum Gripper*". There is, however, no simulation of physical effects after the objects are released again, i.e. the objects would not fall to the ground. Full simulation of physical objects and their interactions is a research issue in itself and far beyond the scoop of research carried out at the Robot Vision Laboratory.

"*Flying object scene*" generates the same room as above but with a flying paper aeroplane instead of the furniture. The plane flies in a circle, varying height and speed sinusodially. If the plane is caught by the "*Vacuum Gripper*" it will stay when released.

"*Transformations*" produces a transformation matrix from translation, rotation, and scale.

"Puma Inverse" calculates the inverse kinematics, translating (if possible) the desired transformation matrix of the end effector to robot joint angles.

"Puma Dynamics" is meant to implement a dynamic model of the robot arm. At the present moment, the module only restrains the maximum speeds with which the joints can rotate. This feature is important, for instance, when using the robot to catch the aeroplane in the flying object scene. Not being able to move instantly from one point in the workspace to another, the system is forced to use some kind of predictor.

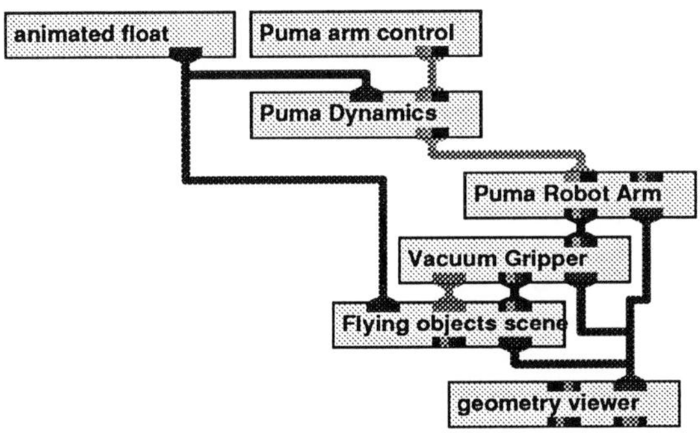

Figure 8: An example of how to use "*Puma Dynamics*". Note, "*Puma arm control*" is a dummy module for illustration purposes.

9. Example of an Experiment

The following experiment setup example show how a complex system can be built in AVS, see figure 9. The experiment concerns tracking of objects on the conveyor belt. Temporal differences is sufficient if there is only one moving object in the scene and the camera is stationary. In this experiment, however, there are a number of moving objects including the robot's own arm, and the camera is saccadicly tracking objects causing temporal differences all over the image.

Of course, the robot is only supposed to track objects of interest without getting distracted by other movements in the scene. A method called Normalized Differential Convolution [5], using certainty fields, provides an elegant means for accomplishing such selective nullification, see figure 12.

The net is driven by "*animated float*", acting as a clock pulse, activating connected modules. Modules not connected to the clock execute as they are called by upstream modules. Note that a module connected both to the "*animated float*"

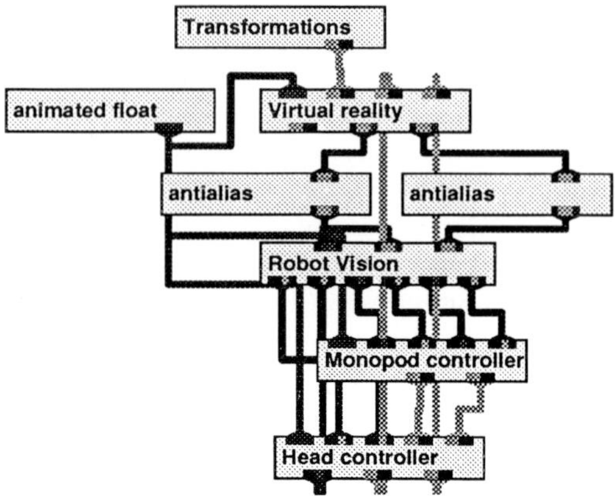

Figure 9: Conveyor belt net. The "*Virtual reality*" macro module produces the images from the stereo camera head, see also figure 10. The "*Robot Vision*" macro module extracts image features used by the gaze control modules, see figure 11. "*Monopod controller*" orients the head according to the potential fields and "*Head controller*" verges the cameras according to the disparity estimates.

module and other upstream modules will normally wait for the upstream modules to finish before executing, but this can be overridden by the programmer.

The robot hand can be positioned and oriented using any module producing a transformation matrix, e.g. "*Transformation*". "*Robot Inverse*" calculates the inverse kinematics, translating (if possible) the desired transformation matrix of the end effector to robot joint angles.

"*Virtual reality*" is actually a macro module containing the sub-network shown in figure 10. It generates the geometric description of the conveyor belt scene, the robot arm, camera head etc. It also renders the images from the stereo camera head. "*Conveyor belt scene*" manages the transformation of manufactured objects and handles robot interaction. Note that it is connected to the clock pulse since it calculates the dynamic behavior of the objects and therefore needs to know the pace of time.

"*Robot Vision*"is also a macro module, see figure 11. It creates a log cartesian fovea representation, phase based disparity estimates, and potential fields around moving objects. "*Generate certainty mask*" and "*Combine certainty mask*" generates and modifies the certainty fields required for the selective data processing. The data flow can be divided into four separate streams which makes it possible to run groups of modules in parallel on four different hosts. In the experiments, four SUN IPX machines where used and "*Robot Fovea Stereo*" was run on a high performance number cruncher, Stardent GS2500.

"*Monopod controller*" orients the head towards the moving object according to

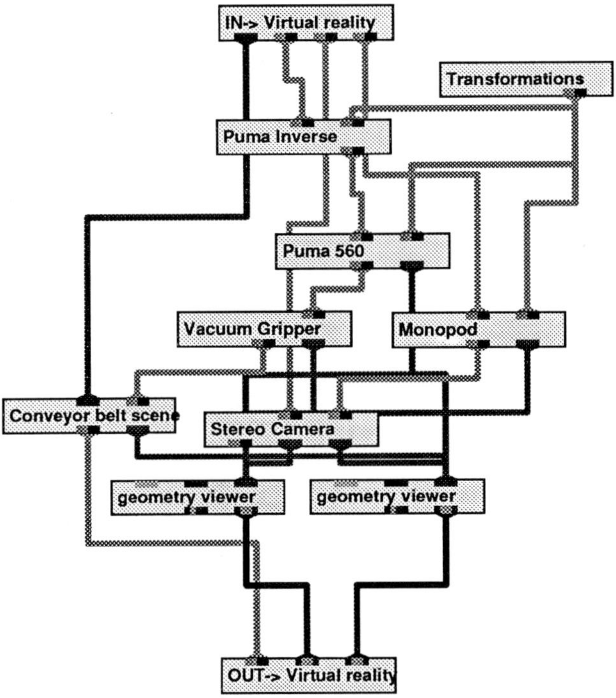

Figure 10: The sub-network corresponding to the "*Virtual reality*" macro module. It generates the geometric description of the conveyor belt scene, the robot arm, camera head etc. It also renders the images from the stereo camera head. "*IN->Virtual reality*" and "*OUT->Virtual reality*" modules provide the interface to the top level net in figure 9.

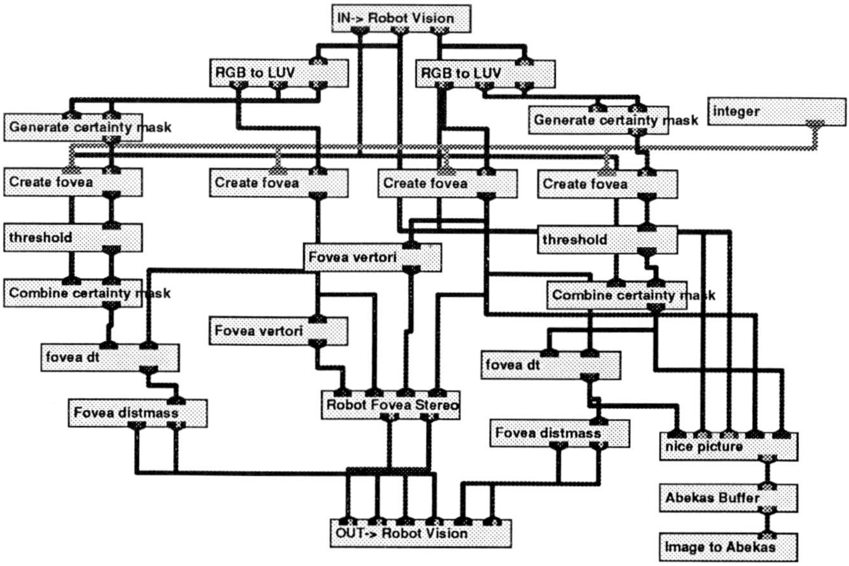

Figure 11: The sub-network corresponding to the "*Robot Vision*" macro module. It generates the log cartesian fovea representation, estimates image disparity, detects moving objects and generates a potential field around the moving objects. The potential field is used by the "*Monopod controller*" to direct the head towards the moving object. Note, the modules in the low right hand corner are for video recording purposes only.

the potential fields using the tilt and pan joints on the monopod. "*Head controller*" verges the cameras to look at the same point according to the disparity estimates.

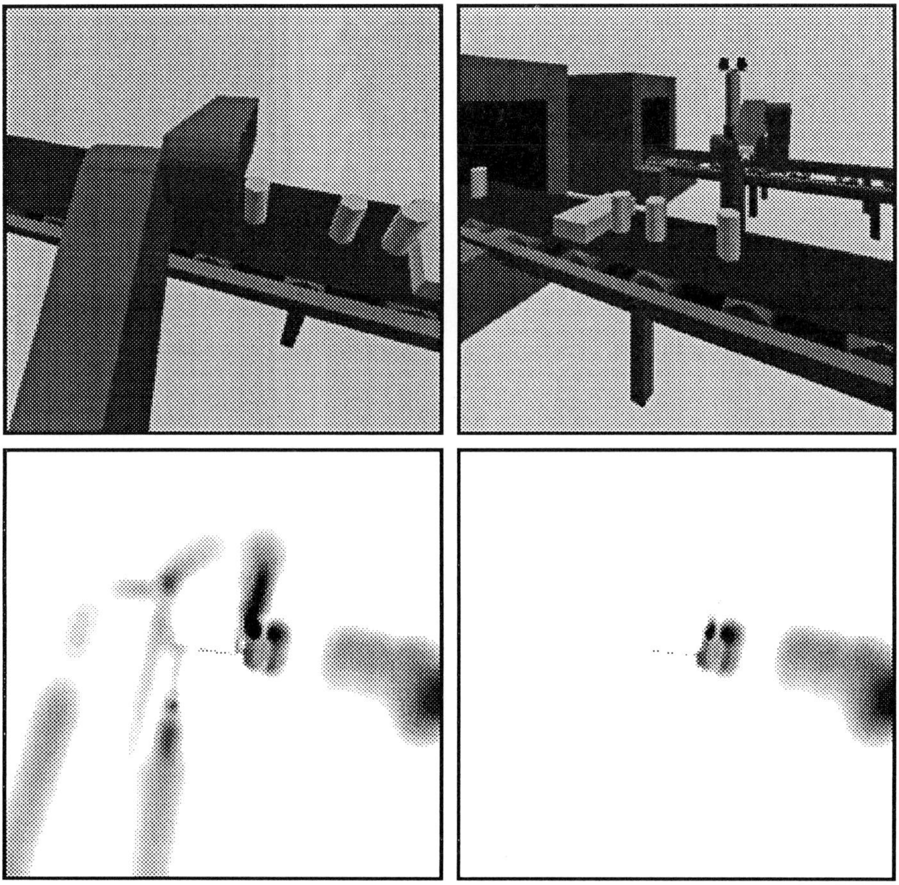

Figure 12: Upper left, the conveyor belt environment. Objects are moving from right to left, while the robot arm is moving back and forth. Upper right, an overview. Lower left, traditional spatio-temporal filtering detecting all motion in the scene. Lower right, Normalized differential convolution and a zero certainty mask covering the robot arm, detecting objects on the belt only.

10. Suggestions for Enhancements and Extensions to Future Releases of AVS

The current release of AVS (release 5) is much more comprehensive and stable than the earliest versions. However, there is still a need for improvements and new features. Following is a list of enhancements and extensions we would like to see in future releases of AVS:

Stop button. If a module is accidently trigged to start, there is no way to stop the execution except throwing it on the hammer, i.e. deleting it. For simple networks and modules the extra work of pulling down and connecting a new instance of the module is negligible. On the other hand, for larger networks with many connections or complicated modules with many parameters to set, it can be very cumbersome to restore the network. A stop button implementing a CTRL–C function is therefore desirable. The fact that a module often starts with deleting its output field and therefore destroys the input field for subsequent modules might be a problem here, but it could perhaps be solved by using temporary memory pointers possibly at the price of memory consumption.

Reinitialize button. If a module is recompiled while an instance of it is present on the workbench it often, but not always, dies and the user can reinitialize it with the current or default parameter values. If the module does not die it will continue to run an old executable and the only way to reinitialize it is to delete it and make a new instance of it. The problem is then again to restore the network. A button that forces a module to reinitialize with the current or default parameters is therefore desirable.

Pipelining facilities. When having continuously changing input data the performance of a network can be enhanced by allowing modules in the same data stream to work simultaneously on data from different time instants, c.f. subsection 4.6. Modules in such a pipeline will have to use double output buffering to avoid memory collisions between new and old data.

Macro module compilation. A macro module does neither take longer nor shorter time to execute than the network it represents. Perhaps time and/or memory could be saved by allowing macro modules to be compiled into one executable.

Data field calculator. The existing field arithmetics modules take two identical input fields and perform one mathematical operation on them, e.g. componentwise multiplication. A module that takes any number of not necessarily identical fields and produces any type of field specified by an expression given by the user would reduce the size of many networks.

Accurate texture mapping. Simulation of moving objects and/or moving cameras can provide test data for image sequence analysis where the ground truth is exactly known. However, the simplifications in the texture mapping introduces noise. The texture seems to float around on the objects. A suggestion is to include the accurate hyperbolic interpolation as a texture mapping option. It may be slower but sometimes accuracy is more important than speed.

Pyramids or multiresolution image fields. Multiresolution image fields can be represented as irregular fields but this is an unsatisfactory solution. Irregular fields require coordinate arrays specifying the location of each point or pixel. This occupies a lot of memory in addition to the fact that pixels might be saved in any order.

Multiresolution fields are more regular than that and a new field type is called for. The current expansion of the Kodak Photo-CD format, which is a resolution pyramid, should be enough to justify the new format.

User libs and object files in automatic Makefiles. The fact that the Makefile is recreated each time the make-button on the "*module generator*" is pressed is called a bug by the AVS user support. This gives hope that this rather annoying behavior is gone when the next release (AVS6) is shipped.

Linked list data format. Dynamically linked lists is a commonly used and versatile form of storing data. Communicating this type of structure between modules is today cumbersome. Some sort of standardized format handling such structures is therefore desired.

Field description. The field structure contains one character pointer for a label and one for a unit corresponding to each vector component in the field. Sometimes, there is also need for a description of the whole field. It might for instance be a history containing the name of the module and parameters with which the field was created. Another use might be a code for how to interpret the data.

Text string editing If the space in a text widget is too small to hold the inserted text the cursor disappears at the right hand edge. The inserted text will be as typed but it is impossible to check for typing errors. For this reason, horizontal scrolling is an absolute need.

Text input editing capabilities without having to retype the whole string is another obvious feature to include.

11. Conclusions

A new system always requires a certain amount of learning and getting accustomed to. This make users uncomfortable and reluctant to change from their old tools. While introducing AVS at the Computer Vision Laboratory, we learned that the height of the getting started threshold could be reduced by simple means. Firstly, it is important that the local file format can be read and written. In the case of AVS this means creating modules for reading and writing the local file format. It is convenient to let these modules convert data to and from an AVS standard format in order to be able to use the full range of modules available. Secondly, it is important to adapt already existing and frequently used operations and visualization methods to the AVS environment. Initially, our laboratory, AVS was used only by a few enthusiasts and did not find a wide spread usage until the modules mentioned above had been created.

The standardized data types and module interface have provided means for the creation of a large common operation library at our laboratory. The graphical presentation of the module libraries helps updating the users about existing modules. This in combination with the in-line help system have increased the exchange of operations and algorithms within our group.

Heavy computations have always been a burden for scientists in the field of image processing and computer vision. The possibility to distribute modules over the computer network is therefore a frequently used and appreciated feature.

On the whole, AVS has been a positive experience. Naturally, there are always features that can be included or enhanced. The "stop button" and the "reinitialize button" should be fairly easy to incorporate, while other extensions might require some fundamental changes of the system structure.

12. Acknowledgments

The support from the Swedish National Board for Technical Development and the Esprit Basic Research Action, Vision as Process, is gratefully acknowledged. The authors also wish to thank Prof. Gösta Granlund and Prof. Hans Knutsson for valuable discussions and suggestions. Finally, we would like to thank Catharina Holmgren for proof-reading this document.

13. References

1. T. A. DeFanti and M. D. Brown, *Visualization: Expanding scientific and engineering research opportunities*, Computer, 22 (1989).

2. D. S. Dyer, *A dataflow toolkit for visualization*, Computer Graphics and Applications, 10 (1990).

3. J. Gallop et al., *Evaluation of visualisation systems; agocg technical report*, tech. rep., Advisory Group on Computer Graphics (AGOCG), c/o Dr Anne Mumford, Computer Centre, Loughborough University of Technology, Loughborough, Leics LE11 3TU, UK, April 1992.

4. A. Kaufman, ed., *Visualization 90*, Los Alamitos, Calif., 1990, IEEE Computer Society Press. Order No. 2083.

5. H. Knutsson and C.-F. Westin, *Normalized and differential convolution: Methods for interpolation and filtering of incomplete and uncertain data*, in Proceedings of CVPR, New York City, USA, June 1993, IEEE.

6. B. H. McCormick, T. A. DeFanti, and M. D. Brown, *Visualization in scientific computing*, Computer Graphics, 21 (1987).

7. Precision Visuals, *Precision Visuals' Workstation Analysis and Visualization Environment (PV-WAVE)*, Boulder, Col., 1988.

8. L. Rosenblum and G. M. Nielson, *Special issue on visualization*, IEEE Computer Graphics and Applications, 11 (1991).

9. C. Upson et al., *The application visualization system: A computational environment for scientific visualization*, Computer Graphics and Applications, 9 (1989).

XITE - X-BASED IMAGE PROCESSING TOOLS AND ENVIRONMENT

TOR LØNNESTAD
Department of Informatics, University of Oslo, Norway and Telemark College, Bø, Norway

and

OTTO MILVANG
Norwegian Computing Center, Oslo, Norway

ABSTRACT

In this paper we present the XITE image processing system. The aim of the system is to be an experimental system. The description of XITE is divided into a five-level hierarchy. The first level is the conceptual definition, the second level is the data structures and basic toolkits, the third level is the image processing routines, the fourth level is image processing programs, and the fifth level is the command interpreter. The system is intended both to give the users a powerful interactive system for image processing, but also, when the experimental part is finished, to give the users all the building blocks they need to realize a stand-alone application. The first will reflect how to work on the two highest levels of the XITE hierarchy, while the latter will reflect how to work on level two and three.

1. Introduction

XITE (pronounced *excite*) stands for "X-based Image processing Tools and Environment". In this paper we will give a brief description of this freeware system, the internal design as well as the external behaviour of the system. Let us first discuss the meaning of the name of the system.

X-based: The system is based on the X11 window system, operating in a UNIX environment. It works with System V as well as BSD versions of UNIX, it is running on a number of different hardware/software platforms. It does not, however, support special hardware like array processors or frame grabbers.

Image Processing: The purpose of the system is to let the user perform image processing in the broad sense, image to image operations as well as image to description operations (image analysis). Currently, though, the image to image operations are outnumbering the image analysis operations. High level operations for machine vision is currently not supported by the system.

Tools and Environment: The system contains a set of tools by means of which the user can create new modules – small routines or large applications – as well as a complete experimental environment ready to perform image processing.

2. The basic design

In this section we will give an overview of the different alternatives available for the design of image processing systems, as well as the design selected for this system.

There are a few books describing and discussing the design of image processing software. In Bolc[1], Chang[3], and Duff[7] a number of existing systems are described thoroughly, while Chang[4] describe the design of such a system in more general terms. Comprehensive reviews of image processing libraries, languages and systems can be found in Denissoff[5], Kulpa[11], Maggiolo[15], Matsuyama[16], Preston[21], Preston[22], and Voss[32].

2.1. Image processing software - an overview

Image processing software systems may roughly be divided into four categories; routine libraries, program libraries, image processing languages, and interactive systems.

2.1.1. Routine libraries

Routine libraries are collections of routines implementing image processing algorithms, following some specified parameter passing format. For every task to be done, a main program has to be written, reading images from disk, calling the appropriate library routines, and writing back to disk. The SPIDER[29] package is one of the largest libraries available, with more than four hundred routines for image and list processing.

2.1.2. Program libraries

A program library can be built on top of a routine library by supplying every routine with a main program. The user interface is a part of each program, and in many cases the different programs behave differently towards the user. A general operating system is used to read user commands and start the appropriate program. Examples of program libraries are Bruner[2], Havens[9], Kirby[10], Landy[12],and Tamura[30].

2.1.3. Languages

A number of programming languages have been designed especially for image processing. Programs have to be written in the language, and compiled and linked before execution, there is no ready system. The languages generally contain special types for images and image related data, and language elements to specify that some operation is to be applied to every pixel in the image. Several languages contain a special operator for two dimensional discrete convolution. Syntactic elements specifying parallel execution is also commonly found in these languages. Several are designed to be run on some special image processing hardware (Gudmundsson[8], Levialdi[13], Vrolijk[33]), but there are also some languages for general computers (Notley[19], Radhakrishnan[23], Uhr[31]).

2.1.4. Interactive systems

Integrated, interactive systems are ready to use systems that contain image processing modules, I/O mechanisms, and a control system interpreting user input or reading commands from a command file, allowing immediate execution of operations specified by the user. Some systems even contain editors, compilers, and access to other operating system services. They are generally built upon a large library of image processing routines, giving unexperienced users a handsome and safe interface to these routines, and eliminating the need to write a program for every task to be done. Khoros[24] is a well known example in this group.

2.1.5. The XITE decision

The interactive systems are the "true" experimental image processing systems, and there are two main design strategies. The system is one large program, where images can remain in memory between the commands, or the system is a program collection tied together by a command interpreter making the system look as one integrated system and not a collection of individual programs.

The I/O overhead needed to read input images and write output images in every program causes a significant overhead compared to systems where intermediate results can remain in memory. Utilizing the UNIX pipe mechanism for intermediate storage will reduce, but not remove, the I/O overhead (Stevens[27]).

Our aim, when designing XITE, was to create an experimental environment for image processing, reducing the effort of using such systems as much as possible. Thus, we had to create an interactive system. In spite of the somewhat lower efficiency of the program collection based systems, this principle was selected due to the higher flexibility of such systems.

An interactive system based on a program collection, where the program collection in its turn is based on a routine collection, constitutes three of the four possible system types in one system. Every algorithm in the XITE system is implemented as a routine where the argument list follows certain conventions. Along with the routine, a main program is written, using common basic routines and conventions for argument handling, making all programs behave in a standard way towards the user, and making the program able to interact with the command interpreter.

2.2. XITE - modules and hierarchies

A complete description of the modules in XITE builds up to a five level hierarchy. The different hierarchies are dependent on the levels below, and in some cases there are internal dependencies on a level, but never upwards. Each level will again be separated in several modules, varying from one to four modules. These are *logical* modules, not neccesarily independent program modules. The five levels are shown in figure 1. As the dependency direction is that higher modules depend on lower ones, we will describe the layers from bottom towards the top.

- Command interpreters
- Image processing applications
- Image processing routines
- Data structures and toolkits
- Conceptual definitions

Figure 1. Five level module description

2.2.1. Conceptual definitions

At this level the system currently contains three modules: a raster module defining the raster image concept used in XITE, a region module defining what we mean with a region in an image, and a display module giving the conceptual basis for the display system. As these modules are pure definitions, there is no software at this level.

The raster module defines an image as a three dimensional structure, but not as a 3D regular array, but generalized to a sequence of 2D arrays. This allows each 2D array (band, channel,..) to have its own size, its own pixel type, and its own displacement and magnification relative to the other bands. As an example, this allows for the pyramid representation to be used. A number of different pixel types are defined, up to double precision complex, with unsigned byte as the most important. An image may also contain any amount of text and other data.

The region module defines a *region* to be a connected set of pixels on a lattice of rectangular pixels. The connectedness may be of type 4 or type 8, corresponding to the number of other pixels that are regarded as "neighbours" of the pixel. An *object* is defined to be a set of regions, regions that, in the users opinion, have some property in common.

The display module will display an image band of arbitrary size and pixel type. Each pixel has a position in a global coordinate system, and the display is a window into this global coordinate system. The window segment may be changed and by this operation zoom and pan through the image.

2.2.2. Data structures and toolkits

At this second level we find four modules: a raster image module, a module for region representation, a display module, and a utility toolkit.

The raster image module (named BIFF) contains in turn a raster image file format, a raster image memory format, and a number of routines handling these formats. The file format is a straight-forward ordering of the information described in the conceptual definition. A single file is used to store an image, starting with an information field specifying sizes etc. Then follows all pixel values, band by band, line by line. Finally the text and other data are placed at the end of the file. The reason for using its own file format is mainly historical, today it would be natural to select a (more) standard one. However, file conversion programs are supplied, and this constitutes no problem.

The data structure used to represent raster images in memory, is the well-known "array of pointers to lines". This structure is described in greater detail in section 3.

The BIFF module contains routines for data transfer between the image file and the memory image. The whole image can be read/written in one call, or specific elements of the image can be transfered separately. There are alse routines for inspecting and modifying the information fields in the image memory data structure.

The region representation is based on the yline data structure. An yline is a structure representing a horizontal, connected sequence of pixels. Ylines at the same vertical position are ordered in a pointer list, and ylines at the different vertical positions are organized in a pointer array. Finally, the different regions in an object are usually put into a pointer list, while the different objects are accessible through another pointer array.

The region module also contains routines for administration of the region data structure. There are routines for conversion between raster and region representation, and for step-by-step construction of a region.

The third module is the display module, which we will focus our attention on and describe in some detail in section 4.

The last module is a general utility module with routines for the handling of program arguments and switches. It also contains a help and information system. Totally, this module gives all the programs a unified appearance to the user.

2.2.3. Image processing routines

This level can be seen as one module, one large bag of algorithms, or it can be split into two, based on whether the routines are raster-based or region-based. A standard way of ordering the arguments is defined, making the argument list of a new routine quite predictable. See section 5 for more details.

2.2.4. Image processing programs

The large majority of the image processing programs are not X-based, they just read input images, process, and write output images. If you want to display the result, the display program is activated. There are however, in addition to the display program, a few programs that are X-based, because they require graphical interaction by the user, typically indicating one or more points in the image. Secondly, the non-X-based programs can be split into raster based and region based ones, summing up to a total of three modules at this level, see section 6.

2.2.5. Command interpreters

To integrate the different programs into one total system, a command interpreter is needed. The system has one command interpreter, a menu based system integrated into the display program. Thus, images being displayed can be piped into image processing programs that are activated through the graphical user interface of the display program. More about this in section 7.

3. The raster image module

3.1. The band and the image

The C language is well suited for algorithms and data structures for image processing (Dobie[6], Piper[20], Shapiro[25]). There are two main methods of organizing a 2D raster in C. The first is the straight-forward 2D array, the second is a 1D array of pointers, where each pointer points to a 1D array of data values. The last method is selected in XITE, as this is a far more flexible structure than the pure 2D array.

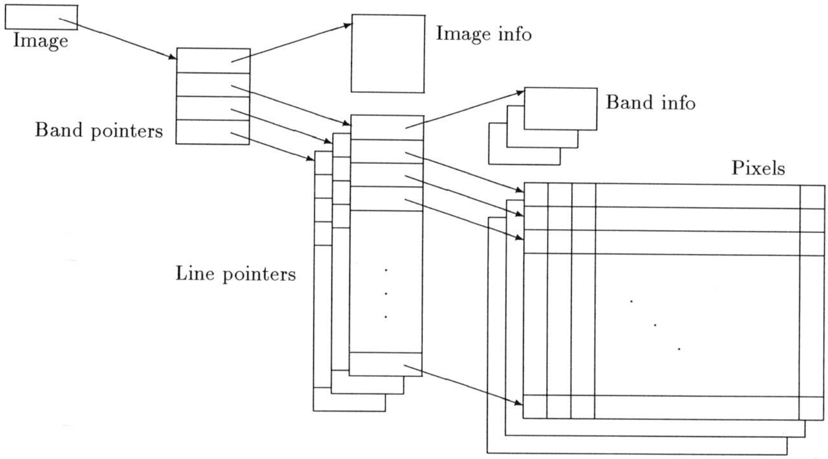

Figure 2. The image structure, 3-band image.

First of all, parameter passing of arrays of arbitrary size is simplified. Using a standard array, the size of the array in all but one direction must be fixed at compile-time. Using the pointer based technique, all arrays are 1D, allowing complete freedom of size. Secondly, the pointer based method has allowed us to include all the other band information into one single data structure, the band data structure. This information includes the size of the band, the pixel type, and the displacement and magnification relative to the other bands. This principle is extended to 3D, defining an array of band pointers, where the first pointer keeps all information related to the whole image. The structure representing a 3-band image is visualized in figure 2.

More than 60 routines support these structures; building the memory structure, transferring data between disk and memory and inspecting/modifying the memory structure. A single "read image" routine opens the image file, builds the whole memory structure, reads all data, and closes the file. Although the system currently supports only up to 3D, the structure is easily extended to more dimensions by creating an array of image pointers and so on.

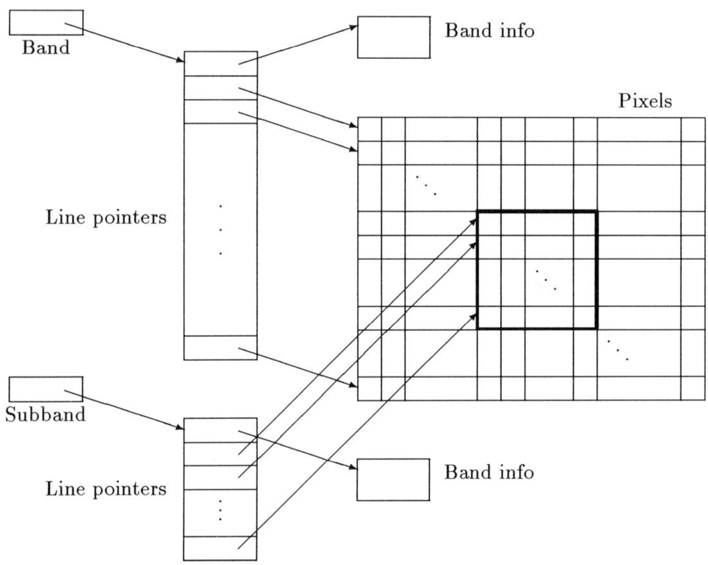

Figure 3. The subband structure

3.2. The subband structure

Keeping the pixel values in the same data structure as the related information, simplifies the parameter passing. Only one parameter has to be passed, the routine can obtain size and other information from the information structure. It also allows another useful structure to be created, the subband. The subband is a virtual band, allowing image processing algorithms to process a rectangular subset of a band, without special attention to the subset coordinates. The subband structure is nearly identical to the band structure, the difference is that the line pointers point somewhere into the middle of another band. The subband structure is illustrated in figure 3.

4. Display tools

In an IPE (Image Processing Environment) you have algorithms and images. A straight forward way to organize your programs is to have one program for each algorithm and one program for visualizing images. This will in general be a good solution, especially when it is integrated with a menu system. The major disadvantage of this type of design is that it is impossible to make a program for visualizing images which covers all your needs. An example may be a warp program, which needs a set of control points as input to the program. If the warp program and the visualization program are two independent programs, the task to keep track of the control points is left to the user. A more user friendly interface is an integrated warp and visualization utility where the user points out the control points with a mouse, and click "warp" when the points are selected. The reason why most IPE's have a separated visualization module is that the visualization is a rather complex task. If we go back to the warp program example, we have two tasks to solve, the visual interface and the warp algorithm. The warp algorithm will, if it is written in a standard language as C, run on all systems from a PC to a CRAY without any modification to the code. The same is not true for the visual interface. Different window systems exist. It seems clear that for workstations, X11[34] is the dominating window system. Within the X11 window system, different methods are used to display images on black/white screens, 256-color screens and 24-bitplanes truecolor screens. We decided to design a toolbox for visualization of images for X11 (Milvang[17]). The interface of this toolbox should provide general methods like "display image", "notify mouse action", "draw in the image", "inspect image (print value)", "zoom/pan image" and so on. The toolbox must of course work for all pixel types.

4.1. An overview

The X11 Window System is an industry-standard software system that allows programmers to develop graphical user interface. One of the most important features of X11 is the device-independent design. X11 provides a toolkit, Xt, for designing "widgets". A widget is an object oriented concept. It is a complex data structure which combines the user interface window with a set of procedures that performs actions on that window, and data needed by these procedures. The procedures may be invoked either by direct procedure call, or by binding actions to the key or mouse. The latter is a convenient and flexible way to design the user interface. All access to the widgets private data are through these procedures. Widgets are organized into classes. In general, a class is a set of things that have similar characteristics. Individual things (objects) are always instances of some class. For example the image widget class described below, defines various attributes common to images: width, height, image data, pixel type, and so on. To display a particular image on the screen, the program creates an instance of an Image. This Image has specific values for all its attributes, for example: width: 512, height: 512, image data: pointer to BIFF-band, pixel type: unsigned byte, and so on. Inheritance is another useful object-oriented concept supported by the Xt toolkit. This means that a class can inherit some or all the characteristics of another class. For example, the ImageOverlayWidget is a subclass of the ImageWidget. In addition to the image attributes, it defines: overlay data, draw color, visible, and so on.

As a tool for visualizing images we have designed an image widget class, an image overlay widget class and a color table pool. The image widget class is divided into four modules: core module, geometry module, pixel module, and display module.

An example:

If you want to display a BIFF-image of float pixels, zoomed in to the upper left corner, on a 8-bitplane bw workstation: Create a new ImageWidget object, set zoom and pan parameters in the geometry model. The ImageWidget knows how to display float pixels on a 256 gray-level screen, see figure 4.

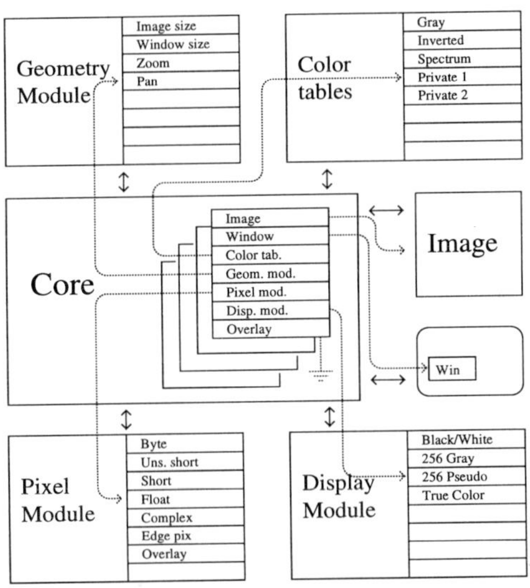

Figure 4. Display an image of float pixels.

The image will be displayed in a window on the screen. The window is a rectangular area of individually addressable dots on the screen. On a black and white screen this dot is set to black or white. On other workstations the dot is set to a shade of gray or to some color. Each dot has a relative position inside the window (w_x, w_y). If the window size is (w_{width}, w_{height}), $0 \leq w_x < w_{width}$ and $0 \leq w_y < w_{height}$, the upper left corner has relative position $(0,0)$. The image is a set of pixels. The description of position for each pixel is hidden inside the widget.

4.2. Pixel representation

A number of pixel types are implemented in the BIFF-format. The ImageWidget must know how to convert the pixel value to a numeric value for gray tone display, a triplet of numeric values for color display, or to a string for reporting the pixel value. It may be necessary to scale and clip the values. Each pixel type in the BIFF-format may have different meaning according to the application. For example in an unsigned byte image the pixel value will represent the intensity for a scanned image, while the pixel value will represent different classes in a classified image. So if you select a point in the image and ask the program to print the pixel value you will expect the pixel value 5 to be printed as "5" for the scanned image and "<Name of class 5>" in the classified image. To the ImageWidget you may specify new (modified) pixel types by defining:

```
typedef int (*resizefunc)();
typedef char * (*printffunc)();
typedef struct pixelstruct
{
char *pixelname;        /* An unique name for the pixel type          */
int  pixelsize;         /* Size of pixel in bytes                     */
resizefunc resize;      /* Implementation details                     */
printffunc printf;      /* Function to convert pixelvalue to string */
int txtlen;             /* Max length of textstring                   */
int scaletype;          /* scale type:  LINEAR or THRESHOLD           */
float scalea, scaleb;   /* lin = a*val+b, thres = val < a ?  0 :  1 */
float minval, maxval;   /* clip values                                */
} pixel;
```

4.3. The display

In the "X window system"[34] display types are divided into 6 classes: StaticGray, GrayScale, StaticColor, PseudoColor, TrueColor and DirectColor. The GrayScale, PseudoColor and DirectColor have read/write look up tables. StaticGray, StaticColor and TrueColor have a fixed look up table. A pixel value indexes the look up table to produce RGB or gray level values. Most common workstations in image processing environments are 8-planes PseudoColor and 1-8 plane StaticGray displays. For each display type there must be a conversion from a gray value or a color triplet to a display value. The widget is responsible to display an image regardless what type of display the user has. In addition to the widget, the programmer also has a pool of color tables available.

4.4. Procedures in the Image Widget

All display commands are sent to the image widget. The image widget will carry out these commands by itself or by using virtual procedures from the other modules. The methods in the image are:

Create display object:
Initialize the data structure for the object and tell the window system to create a screen window. Read screen characteristics and select a method for image visualization according to the screen type. Modify the magnification if necessary.

Destroy display object:
Deallocate data structures and tell the window system to remove the screen window.

Redisplay display object:
Redisplay the entire window or a subarea of the window.

Set new window size:
In most window systems the user may resize windows on the screen.

Get current zoom and pan parameters:
Return the currently used parameters.

Set new zoom and pan parameters:
Set new zoom pan parameters. This procedure is invoked directly or as a key and mouse action. It is very convenient to bind the mouse to these actions:

```
Shift<Key>Right:   zoom_pan(+0, +16, +)
<Key>Left:         zoom_pan(+0, -4, 0)
Ctrl <Btn1Down>:   zoom_pan(*2, MousePosC, MousePosC)
Ctrl <Btn2Down>:   zoom_pan(1, MousePosC, MousePosC)
```

The four lines above is an example on how you bind some key/mouse events to the actions: pan 16 pixels right, pan 4 pixels up, zoom in by a factor 2 with mouse position in the center, and set zoom factor 1:1.

Common geometry:
If "Common geometry" is set, all images on the screen are forced to have the same geometry parameters. This is useful when comparing two or more images.

Convert window coordinates to image coordinates:
According to the current settings, this routine will convert window coordinates to image coordinates.

Convert image coordinates to window coordinates:
Depending on the magnification, one pixel may be mapped to zero, one or many dots in the window. If one pixel maps one dot, the coordinates of the dot are returned, otherwise a parameter to this procedure tells how to pick the best position (left or right candidate).

Convert window coordinates to pixel address:
A pixel may be of any size and type. This routine will only give a pointer to the pixel value. The application is responsible for using the address correctly.

Convert pixel to unsigned short:
This method is used to display gray-level images. Each pixel is converted to an integer in the range 0..65535.

Convert pixel to triplet of unsigned short:
This method is used to display color images. The triplet represents the three values in the *RGB* color system.

Convert pixel to string:
This method will return a string where the pixel value is converted to ascii text. This gives the display system a method to report pixel values.

Set parameters for the pixel representation:
Normally you will get the current descriptor, modify it and write it back. This gives an easy way for example to set only a new print function to the pixel structure.

Print string into display:
The font size and text length from the pixel structure is used to calculate the space used by the text. If an image is magnified enough, each pixel will cover an area big enough to display the textual value of the pixel.

Notify events:
Event notification may be set for different types of events: Mouse move, mouse button down, mouse button up, mouse event in combination with <SHIFT> or <CTRL>, resize window, focus change, and so on. For each event a user defined procedure is invoked with an ImageCallback structure as parameter:

```
typedef struct
{
int reason;
XEvent *event;
struct imagestruct *img;
struct bandstruct *band;
int x, y;
ImageData data;
pixel *pixtype;
float zoom, magnification;
int xpan, ypan;
} ImageCallbackRec, *ImageCallback;
```

The application may use this data to inform the user about cursor position, pixel value, which image segment is used, an so on. An example on how to report cursor position and pixel value is:

```
static void Button(ImageWidget wid,
caddr_t dummy, ImageCallback button)
{
char buf[MAXLEN];
(button->pixtype)->printf(buf, button->data);
Message_pos(button->x, button->y, buf);
}
```

Regardless of what pixel type that is displayed in the window, this procedure will print the correct value.

4.5. Overlays

As a subclass of ImageWidget, we have defined the ImageOverlayWidget. The overlay is represented by an unsigned byte band of the same size as the image. If the overlay value is zero, it is transparent, otherwise it will be displayed as a colorpoint. The overlay will have the same zoom and pan parameters as the underlying image.

The system is designed to make it possible to draw in the overlay band with the mouse. The most important actions for drawing are:

draw(origo[, on/off])
on: Set new origo in the drawing module to mouse position. If the drawing state was "no origo" this point is set as the start of a new path.
off: Unset origo in the drawing module.

draw(point [,size])
Draw a point, or cross of specified size, at mouse position.

draw(line [,close])
Draw a line from origo to mouse position. If the parameter close is used, draw a line from origo to beginning of path.

draw(fill/clear [, arguments])
Fill/clear region. The arguments are used to specify neighborhood relations, fill type and so on.

draw(graphic, on/off/toggle)
Turn display of overlay plane on/off.

draw(color, value)
Select a draw value. Must be in the range 0..255.

These commands may be merged to powerful drawing commands. For example in a region drawing program:

```
None <Btn1Up>:       draw(line) draw(origo)
None <Btn1Down>:     draw(origo,on) draw(point)
None <Btn1Motion>:   draw(line) draw(origo)
```

Since resources may be set by all users, a user may easily change the behavior of the application to draw straight lines. In the following code double-click is used to close the path.

```
<FocusIn>:           draw(origo,off)
None <Btn1Down>:     draw(line) draw(origo,on)
None <Btn1Up>(2):  draw(line,close) draw(origo, off)
```

4.6. Color tables

In image processing labs, color workstations are popular. Most of these are able to display an image with up to 256 simultaneous colors. If a workstation has only one hardware color table, and you set up the display to use all colors on one image, then the screen will flicker a lot when you move the mouse around. The flickering is caused by the change of color table for different applications. We have solved this problem by using a color table pool. The image widget will normally use a 128 entries color table, and if it has an overlay it will allocate another 32 entries. A standard grayscale and 32 overlay colors are allocated from the default color table (if possible). This color table is shared by all images, so when moving the mouse between these there will be no flickering. In addition to the standard color table we offer a set of 8 different sharable color tables. Applications using the color table pool will, when they start up, search for a file "xite_color_tables" (or a file specified on the command line). This file will tell the program which color tables to load. Below is an example of private "xite_color_tables" files:

```
!  Sample color tables file.  (!  is used for comments)
!
^$XSHOWPATH/xshow_color tables  !  Read the system color table_file
^/local/classify/color tables   !  Add color tables for classification
!
:Standard color tables          !  Add some private color tables
mywhite.col
myblack.col
!
:Overlay color tables           !  Add a private overlay table
myoverlay.col
```

The strength of this concept is obvious. People working together on a project may share some color tables, so they are sure to use a common color coding for their image presentation. Regardless of what X-application you are running you have available the same set of color tables.

4.7. Histograms

Since histograms and greylevel adjustment are important in image processing, we have designed a histogram tool which may be invoked by a single procedure call, see figure 5. In this toolbox you may inspect the histogram and the cumulative histogram, and if the display supports color tables you can interactively modify the color table. The new color table may be added to the color table pool. If the display doesn't support color tables, it is still possible to use the histogram tools. Instead of modifying the color table, the whole image is recalculated and redisplayed.

4.8. Color slicing

An interactive color slicing module is also available if the display supports color tables, see figure 5. Colors may be specified in RGB, or IHS. As for the histogram, the new color table may be added to the color table pool.

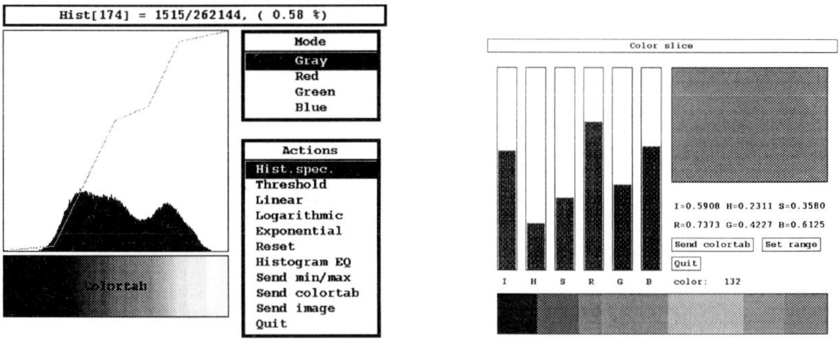

Figure 5. The Histogram and Color slice tools.

5. Image processing routines

Instead of going into details about any particular routine, we will here describe the general structure of an image processing routine, and then give a brief overview of the contents of the image processing routine library.

5.1. Routine structure

Most image processing algorithms are defined to work on a two dimensional image. To follow the principle of building small, elementary program units that can be combined or iterated into more advanced functionality, the XITE routines also work on the two dimensional raster, denoted "band" in this system. Thus, it is up to the main program to iterate over all bands if desired. In a few cases the algorithm requires an image, an obvious example is routines for multi-spectral classification. On volume images we need 3D filters, these will also require 3D arguments.

The parameter sequence is ordered in 1) input bands, 2) output bands 3) other input arguments, and 4) other output arguments. The functions return a status value, where 0 indicates success, and different other values indicate different kinds of errors. The routine is a separate module without any global variables, but it will report information, warnings and errors to a message system. This system may be tailored by the main program, and by the user, choosing if and how such messages should be reported.

5.2. Contents of the library

To give an impression of the contents and profile of the library, approximate numbers of algorithms within several different classes of problems are given. In addition there is a large number of routines not fitting into any of these groups.

Orthogonal transforms : 22 routines
Histogram operations : 11 routines
Edge detection : 6 routines
Noise reduction : 15 routines
Texture analysis : 25 routines
Arithmetic operations : 8 routines
Logical operations : 25 routines
Morphology : 4 routines
Thresholding : 20 routines
Reorganization: 5 routines
Classification : 2 routines
Region operations: 20 routines

6. Image processing programs

The image processing programs can be separated into three groups, the few ones that are X-based into one group, and the others are separated into raster based and region based programs. In this section, we shall first have a look at the use of the non-X-based programs, being used as a program collection. How are they activated from the shell, how do they fit together. Then we shall have a closer look at some of the X-based programs.

6.1. The program collection

In general, we have a collection of programs, where each program is connected to one image processing routine. The program reads input arguments, calls the image processing routine, and writes the output arguments. The basic utility toolkit is used at all stages here, giving all programs some common behaviour regarding parameter handling, error messages and help system. As an example, all programs know how to handle switches like "-help", "-man", "-whatis" and "-verbose". All programs append a line to the text field of the processed images, with information about what program has been run, when, and with what arguments. Thus, all images contain their processing history for later examination.

To perform a simple median filtering, the command is as could be expected:

```
$ median <input> <output> <filter size>
```

However, when used in a piping sequence, the syntax is not what might be expected. If an XITE program is activated without arguments, it will generally give the usage and terminate, instead of waiting forever for the user to supply the image pixel by pixel. To indicate piping, the special file name "-" is used.

A simple example will illustrate the use. To obtain significant edges in an image, it is median filtered and also mean filtered. Since the median program and the mean program should have the same image as input, biffcopy is used to duplicate the image. The absolute difference between these results is thresholded at pixel level 10, and displayed with the xshow program.

```
$ biffcopy <input> - - |\
( median - - 5 ; mean - - 5) | \
absDiff - - - | \
threshold - - 10 | \
xshow -
```

Based on this system, powerful processing sequences can be described, and eventually saved in a shell script for later use. A few programs are particularly well suited for script usage. The "biffinfo" program extracts information about the image, by means of different switches it can extract sizes, pixeltypes, the text field (the processing history) and so on. The "statistics" program reports statistical features extracted from the pixel values, like mean value, median value, maximium and minimum value, and standard deviation.

6.2. Region analysis

In the previous section we described tools for region analysis. We also described tools for drawing/editing an overlay image. In the region analysis program "xregion" these tools are combined to a powerful interactive image analysis program. Input to the program are the image to be analyzed and an optional region image (overlay image), which might be the result of a previous segmentation. A new region image will be created if it is not a parameter to the program. The drawing tools are used to draw/edit regions. The overlay image may be sent to the "regionAnalyse" program to extract features or the "statistics" program to extract some simple statistics. Toggle buttons are pressed to indicate which features to be extracted.

Another task for this program is to draw training masks in for example satellite data to establish a class description base for supervised classification. Figure 6 shows the letters O V E R L A Y drawn into the image. Each letter has a different overlay value. The result of the analysis shows parameters from the background, the 7 letters, and the holes from O, R, and A.

6.3. Sequence browser

A program, "xmovie", is used to view a sequence of images, such as slices through a 3D object or time series of 2D images. The only limitation is that the images to be displayed all have the same size and pixel type. If you use zoom or pan, all the images in the series are zoomed and panned with the same parameters. It is possible to manually step one frame forward and backward, or step forward with a fixed time interval. For 256x256 images the max speed is 15 images/sec on a DecStation 5000/200.

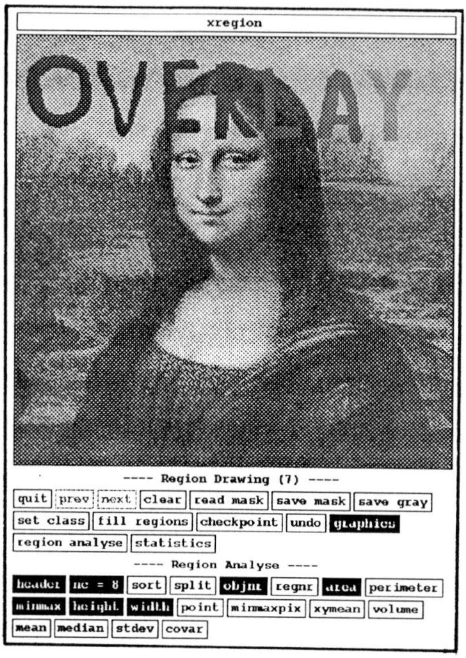


```
***** Job 13545: regionAnalyse *****
# OBJNR   AREA    XMIN   YMIN   XMAX   YMAX   HEIGHT WIDTH
#----------------------------------------------------------
     0   243409      1      1    512    512    512    512
     1     3237     12     25     89    111     87     78
     3     2307    173     29    218    116     88     46
     2     1742     96     31    154    121     91     59
     4     2770    235     31    284    122     92     50
     5     1419    290     37    336    121     85     47
     7     1832    410     37    462    123     87     53
     6     2666    352     39    404    125     87     53
     0     2219     27     40     76     94     55     50
     0      302    250     46    266     66     21     17
     0      241    375     63    387     92     30     13
***** Job terminated *****
```

Figure 6. The xregion program.

7. An image viewer and command interpreter

This is the most important tool for visualization in the software package. The program, xshow, is able to display a number of images only restricted by the computer limitations. The program takes advantages of the display tools such as zoom, pan, printing of pixel values, switching overlays on and off, and so on.

The xshow program may display several images. Each image band has its own window. Each image in the system has a pointer to a colortable. A colortable may be shared by many images or may be private. A histogram tool and a color-slice tool will create a private colortable for an image, and this colortable may be modified by interaction with the user.

From the xshow main window it is possible to start an IPP (image processing program). When an IPP is started, the user uses the mouse to click out the image(s) he wants to process. The image sent to the process may be a Region Of Interest (subband), a single band, or the entire image. The pipe mechanism described previously is used to feed the IPP with data. The output image(s) from the IPP is sent by a pipe back to the xshow program which will display the result.

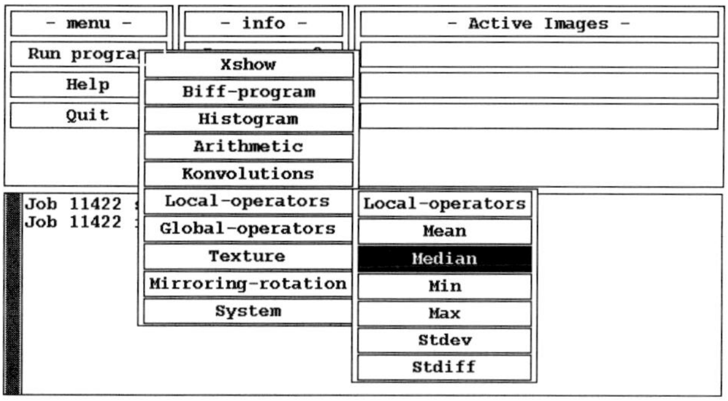

Figure 7. Example of the program menu.

The user may design his own menu. As for the colortables you may include system-files. Below is a fragment of the system menu-file used to create the menu shown in figure 7.

```
!
!  Main menu - Xshow
!  Some special arguments:
!       <infile>    - Read input image from xshow by mouse command
!       <outfile>   - Send output image to xshow
!       ?writefile  - Standard dialog (Ask for outfilename)
!       ?readfile   - Standard dialog (Ask for infilename)
!       ?dialog:xx  - Fetch argument(s) from dialog xx
!
:Xshow                  !  Menu name
+Biff-program           !  Entry is a menu
:          :                        !  Some lines skipped
+Local-operators        !  Menu = Local operators
:          :                        !  Some lines skipped
!
:Local-operators
!Name                   ; unixcommands <args>
Mean                   ; mean  <infile> <outfile> ?dialog:size
Median                 ; median <infile> <outfile> ?dialog:size
:                                         :
!
!  ---- Dialogs ------
!
?size                        !  name
Enter "Window size"\
or    "width" "height" !  text (2 lines)
3 3                        !  default answer
```

When the "Median" entry is pressed the median command is about to be executed. From the manual page for median we read:

```
SYNTAX
median [-frac fractile] [-mask maskfile] [-mirror]
<inimage> <outimage> [<dx> [<dy>]]
```

The dialog "size" is started. This dialog will return a text containing the user supplied argument(s), eventually the default arguments. The median program is forked out as a subprocess from xshow as the command:

```
median -<n> -<m> <size>
```

<n> and <m> are channel numbers. When the user clicks the mouse in an image on the screen, this image (the entire image, the band, or a subband of it) is piped into channel <n>, and when the program has calculated the result this is piped out on channel <m>. "size" may contain more than one parameter, so the answer "5 7" will set the dx-parameter to 5 and dy-parameter to 7 (width and height of filter window). So the reader may ask: Is it legal to set "size" to " 5 5 -frac 0.2"? The answer is yes. There is no syntax check on the dialog answer, so all input which will form a valid syntax to the program is legal. If you try to enter "size" as "3 5 7" this is a valid answer for the dialog, but since it will form an illegal syntax, the median program will only respond with a usage message.

The xshow control window gives information about existing images and running jobs. Error messages from the running program will also be written to this window. After running median with the parameters above, the message log may look like:

```
Job 14191 started  - median <infile> <outfile> 5 7
Job 14191 finished - median <infile> <outfile> 5 7
Job 14194 started  - median <infile> <outfile> 5 5 -frac 0.2
Job 14194 finished - median <infile> <outfile> 5 5 -frac 0.2
Job 14204 started  - median <infile> <outfile> 3 5 7
Job 14204 message  - Usage:  median [-frac fractile] [-mirror]
[-mask mask] <inimage> <outimage> [<dx> [<dy>]]
Job 14204 finished - median <infile> <outfile> 3 5 7
```

8. Applications

In this section we will show two areas where the xite system is used to build applications for operational usage. The first application is from remote sensing[26], the second from medical imaging[28].

8.1. Snow-cover monitoring

Knowledge of the amount of snow left in a drainage basin is important to power companies. In Norway, a substantial part of the drainage basins is located in high mountain areas where more than 50% of the precipitation falls as snow. The amount and distribution of the snow is crucial for planning of the power production. For several years, Norwegian Computing Center has provided Statkraft (the largest Norwegian hydroelectric power company) with analysis of images from the NOAA AVHRR sensors in order to determine the snow cover throughout the melting season. The software system used in this project was based on the XITE-library and the user interface has the image-widget as the major component, see figure 8. The system is now in operational use at Statkraft. Geometrically corrected NOAA/AVHRR images are acquired from Tromsø Satellite Station. The first step in the analysis is to use XITE-programs to convert the NOAA-images to BIFF-format and to convert it to byte-images. The next step is a classification system. The display module in XITE is a backbone for interactively attaining a correct parameter set for the classification. The last step is the generation of three different map-types. The colortable pool is used to support correct color codes in the maps, and the BIFF to PostScript conversion utility is used for printer output. The classification algorithm itself is written for this application and is not a part of the XITE-system, even if some of the utilities (such as the history record) is used.

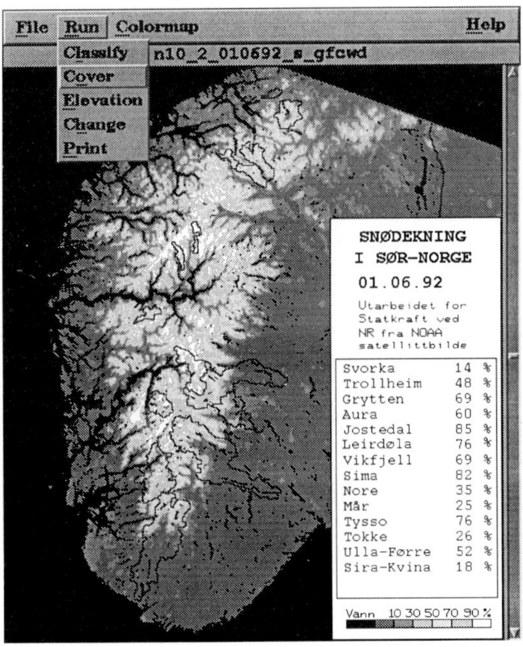

Figure 8. Snow cower in South Norway.

The user interface is based on Motif X Windows system under UNIX, and provides user interaction through menues combined with display of image data and map products on the screen. Sensible default parameter values combined with options for automatic parameter estimation makes very efficient use possible.

8.2. Volume measurements of brain parenchyma

The aim of this project is to develop and test methods for performing semi-automatic segmentation and volumetric measurements of brain parenchyma and cerbospinal fluid in 2D and 3D MRI images. In this project XITE is used for visualization of sequences of slices from the brain. The region analysis tool is used to estimate brain parenchyma and cerbospinal fluid.

9. Discussion and conclusion

The XITE system has been presented in general terms, and some parts of the system have been described in greater detail. The XITE project has been going on for about nine years with varying effort. It started with the file format definition in 1984-85, on a VAX/VMS system, implemented in Pascal. In 1987 it was rewritten in C for a UNIX environment, later on an X-based display program was developed, and it slowly grew to a complete system. Currently the system contains nearly 200 programs, a slightly higher number of image processing routines, and a large number of basic utility routines.

9.1. Experience with the system

The system has been used these years at the University of Oslo, as an experimental tool in undergraduate courses in image processing, and as a research tool for graduate students and for the staff. To fulfil the educational requirements, nearly all the algorithms described in Niblack[18] have been implemented. Through the last years some 40 other institutions world wide have started using XITE, both within academic institutions and applied research. This description and discussion is thus based on experience and comments from a large number of people for several years. We are grateful for all contributions.

9.2. Availability

XITE is a freeware system, latest version is available from the University of Oslo, Norway, with anonymous ftp from ftp.ifi.uio.no (currently a synonym for 129.240.82.2). Look for the files "README", "install" and "xite.tar.Z" in the blab/xite directory. Then have a look at the README file for information about the installation. The tar-file constitutes about 2 MB, and you should have at least 20 MB (for SPARC code, a lot more for DecStation) free for the installation. There is a mailing list for the exchange of information about XITE, this list is named "xite@ifi.uio.no". To join the list, or if you have any kind of questions to the authors, mail to "xite-request@ifi.uio.no".

9.3. Room for improvements

There are several aspects of the system that has not yet received the attention and resources they deserve. In other words, there are plenty of room for improvements. Some elements are:

The system is not as complete as could be wished on some fields. Several important areas are empty, or nearly empty, like high level computer vision and plotting/3D-visualization.

Lønnestad[14] presents a taxonomy of interaction levels and corresponding languages, and four groups are identified. Command languages (including graphical languages) are used to give individual commands for immediate execution. Macro languages are used for storage and later execution of command sequences. Interpretative languages are typically used for fast prototyping of image processing algorithms. And finally compiled languages are used for later execution in high speed. Graphical languages are suitable at the two first levels, typically menu languages and visual programming languages respectively, but at the two last levels traditional high level languages, eventually object oriented languages, seem to have no competition. The ideal system should offer interaction at all four levels, and preferably with the same language at all four levels, in addition to optional graphical languages at the two first levels. XITE, as well as all other existing systems to our knowledge, fails to fulfil this requirement. Some systems support all levels, but with different languages. Other systems support the same language at up to three levels. The XITE system offers shell commands and menu commands at the command level, shell scripts at the macro level, no language at the interpreter level, and C at the compiler level. Extending the system to fulfil this requirement would be an interesting project.

XITE does not support visual programming, which seems to be an increasingly popular feature. Insted of developing a new such system, the XITE programs could be adapted to the conventions of an existing system like Khoros[24].

Currently, XITE only supports up to 3D data. This should be extended at least to 5D, but preferably to any-D.

Although the system supports a number of pixel types, it is up to the person implementing an algorithm to take the effort of implementing it on all pixel types. Most of the algorithms in XITE support only unsigned byte pixels, but direct conversion between all pixel types is supported.

9.4. Advantages of this system

According to our experience so far, there are also a number of aspects where the system seems to be at least competitive with the alternatives.

For a beginner, the system is very easy to use. Default arguments are used systematically, to reduce the number of parameters that has to be specified. The threshold to cross for an unexperienced user is very low.

Although flexibility is more important than efficiency, the system has proven to be quite efficient as well. Some examples are 0.3 sec for the sobel operator, 0.5 sec for general 3x3 convolution, and 2.4 sec for a 11x11 median filter, all measured on a 512x512 byte image on a DecStation 5000/240. Times are user times, including reading image from file, performing the algorithm, and writing the result back to file.

In some areas, the system has a large number of algorithms. Some examples are noice reduction and thresholding algorithms. The region system is also quite comprehensive.

The system is easily modified, even by each individual user. A first example is the addition of new algorithms, which can be user-individual, or installed in the common system. A second example is modification of the menu hierarchy, again modifications can be done by each individual, or by the local administrator and influence all the users.

Finally the system contains powerful basic tools for the construction of new complete applications. These tools are at the second and third level of the design description: raster image handling routines, region handling routines, utilities for help, information, error and argument handling, the display module, and the raster- and region based image processing routines.

The XITE system satisfy most of our demands for an experimental image processing system, and has proven to be an efficient tool. However, we are continously working to improve the basic system, as well as enlarging the set of implemented algorithms.

References

1. Bolc, L. and Kulpa, Z.(eds) "Digital Image Processing Systems" *Lecture Notes in Computer Science* **109**, Springer Verlag, 1981.
2. Bruner, J.D. and Reeves, A.P. "An Image Processing System with Computer Network Distribution Capabilities" *Proc. of Prip. 82 IEEE Comp. Society Conf. on Pattern Recognition and Image Analysis*, Las Vegas, 1982, pp 447-450.
3. Chang, S.K. and Fu, K.S.(editors) "Pictorial Information Systems" *Lecture Notes in Computer Science* **80**, Springer Verlag, 1980.
4. Chang, S.K. *Principles of Pictorial Information Systems Design* Prentice-Hall, 1989.
5. Denissoff, G. and Rämö, E. "A Survey of Digital Image Processing Systems" *Proc. of the 2. Scand. Conf. on Image Analysis*, Helsinki, Finland, 1981, pp 39-46.
6. Dobie, M.R. and Lewis, P.H. "Data structures for image processing in C" *Pattern Recognition Letters*, **8**, 1991, pp 457-466.
7. Duff, M.J.B. and Levialdi, S.(eds) *Languages and Architectures for Image Processing* Academic Press, 1981.
8. Gudmundsson, B. "Overview of the High-Level Language for PICAP" in Bolc[1], pp 147-156.
9. Havens, W. "PIPS: A Portable Image Processing System" *Proc. of the 8th Intern. Conf. on Pattern Recognition*, Paris, 1986, pp 125-135.

10. Kirby, R.L., Notley, D.P., Rosenfeld, A. and Weber, J.C. "TIPS: Transportable Image Processing Software" *TR-1222*, Comp. Vision Lab., Comp. Sc. Center, Univ. of Maryland, 1983.
11. Kulpa, Z. "Universal Digital Image Processing Systems in Europe – a Comparative Survey" in Bolc[1], pp 1-20.
12. Landy, M.S., Cohen, Y. and Sperling, G. "HIPS: A UNIX-Based Image Processing System" *Computer Vision, Graphics, and Image Processing* **25**, 1984, pp 331-347.
13. Levialdi, S., Maggiolo-Schettini, A., Napoli, M., Tortora, G. and Uccella, G. "On the Design and Implementation of PIXAL, a Language for Image Processing" in Duff[7], pp 89-98.
14. Lønnestad, T. *Design and Implementation of an Image Processing Operating System* Ph.D. thesis, Dept. of Informatics, University of Oslo, 1991.
15. Maggiolo-Schettini, A. "Comparing som High-level Languages for Image Processing" in Duff[7], pp 157-164.
16. Matsuyama, T. *A Survey of Image Processing and Computer Vision Software systems on Workstations* Report from TC5, International Association for Pattern Recognition, 1992.
17. Milvang, O. and Lønnestad, T. "An image widget for image processing" *The X Journal*, **1**, No 2, 1991, pp 62-67.
18. Niblack, W. *An Introduction to Digital Image Processing* Prentice-Hall, 1986.
19. Notley, D.P. "LOP: A Language for Image Processing" *TR-CAR-19*, Center for Autom. Research, Univ. of Maryland, 1983.
20. Piper, J. and Rutovitz, D. "Data Structures for Image Processing in a C Language and UNIX Environment" *Pattern Recognition Letters* **3**, North-Holland, 1985, pp 119-129.
21. Preston, K. "Image Processing Software - A Survey" in Kanal, L.N. and Rosenfeld, A.(eds): *Progress in Pattern Recognition* North-Holland 1981, pp 123-148.
22. Preston, K. "Progress in Image Processing Languages" in Duff, M.J.B.(ed): *Computing Structures for Image Processing* Academic Press, 1983, pp 195-211.
23. Radhakrishnan, T., Barrera, R., Guzm an, A. and Jinich, A. "Design of a High-Level Language (L) for Image Processing" in Duff[7], pp 25-51.
24. Rasure, J, Argiro, D., Sauer, T., and Williams, C. "A Visual Language and Software Development Environment for Image Processing" *International Journal of Imaging Systems and Technology*, 1991.
25. Shapiro, L.G. "Data structures for picture processing: A Survey" *Computer graphics and image processing*, **11**, 1979, pp 162-184.
26. Solberg, R. and Andersen, T. "Operational Monitoring Of Snow Coverage And Soil Erotion In Norway" *International Symposium on Operationalization of Remote Sensing*, 19-23 April 1993, ITC, Enschende, The Netherlands.
27. Stevens, W.R. and Hunt, B. R. "Software Pipelines in Image Processing" *Computer Graphics and Image Processing* **20**, 1982, pp 90-95.

28. Storvik, G. and Lundervold, A. "Segmentation of Brain Parenchyma and CTF in Multispectral MR Images of the Head" *Proceedings of the 8th Scandinavian Conference on Image Analysis* Tromsø 1993, pp 3-4.
29. Tamura, H., Sakane, S., Tomita, F., Yokoya, N., Kaneko, M. and Sakaue, K. "Design and Implementation of SPIDER – A Transportable Image Processing Software Package" *Computer Vision, Graphics and Image Processing* **23**, pp 273-294.
30. Tamura, H., Sato, H., Yamamoto, H., Okazaki, H. and Kawai, T. "The Software Architecture of an Image Processing WorkStation" *Proc. of the 6. Scand. Conf. on Image Analysis*, Oulu, 1989, pp 763-769.
31. Uhr, L. "A Language for Parallel Processing of Arrays, Embedded in PASCAL" in Duff[7], pp 53-87.
32. Voss, K., Hufnagl, P. and Klette, R. "Interactive Software Systems for Computer Vision" in Kanal, L.N. and Rosenfeld, A.(eds): *Progress in Pattern Recognintion* North-Holland, 1981, pp 57-78.
33. Vrolijk, J., Pearson, P.L. and Ploem, J.S. "TAL: An Interpretative Language for the Leyden Television Analysis System" in Duff[7], pp 125-137.
34. Young, D.A. *The X Window System, Programming and Applications with Xt OSF/Motif edition,* Prentice Hall, 1990.

LABOIMAGE: AN EXPERIMENTAL WINDOW-BASED ENVIRONMENT FOR TEACHING AND RESEARCH IN IMAGE ANALYSIS

ALAIN JACOT-DESCOMBES

and

THIERRY PUN

Computer Vision Group, Computing Science Center, University of Geneva
24 rue du Général-Dufour, 1211 Geneva 4, Switzerland
email: [jacot,pun]@cui.unige.ch

ABSTRACT

LaboImage provides teachers and scientists with general purpose as well as specific processing families and tools in a highly interactive and experimental environment. The current software reflects half a decade of development and experiences. Throughout this paper, the initial need for such a software is presented as well as its evolution. The current features are listed; also described is the LaboImage interface design and how users interact with the software. The implementation aspects are then detailed, concerning data structures as well as algorithms and interfaces. The data file and descriptor file format used for storing images is described. The organization in memory of multiple data such as images, vectors and macros is presented, as well as the source code organization. A clear separation between algorithmic and interface code has proven to be very important, in order to allow easy extensions of the basic system. Finally, some aspects that distinguish LaboImage from other imaging softwares are discussed.

1. Introduction

The initial discussions from which LaboImage originated were motivated by the need of an experimental environment for students taking an image processing and analysis course. The lack of such a platform implied the need for students to spend a significant amount of time to program classical processing algorithms, rather than understanding and using them. The project began in summer 1987; the initial specification was to construct a window-based environment for manipulating images and to make transparent calls to processing routines from libraries such as Spider [14] or, alternatively, from self-written routines. The first version was used by students during the academic year 1988-1989 [6]. This version was providing students with standard image processing and manipulation methods, such as display, enhancement, filtering and mathematical transforms [4].

The advantage of such a platform is also to offer the possibility of integrating other works performed in the Computer Science department, such as a mathematical morphology package or a look-up table editor. Such extensions make LaboImage constantly evolving, providing users with more possibilities than the standard ones that can be now found in various, general imaging packages. Specific applications have also been separately developed for responding to particular needs, especially in biomedical imaging: examples are softwares for digital analysis of superimposable dental radiographs [3] or 1-D gel electrophoresis analysis [9].

LaboImage tries to provide as general a platform as possible for various scientific investigations, while still offering a few selected, specialized tools for more specific tasks. It provides students or scientists with a large choice of processing procedures able to solve various classes of different problems in a user-friendly environment. LaboImage is typically suited for exploratory research, where one tries to compare different methods, by allowing rapid determination and evaluation of solutions to a given problem. This software has been distributed in various computer centers and universities over the world, and is also used as a platform for an image analysis course given for computer science students at University of Geneva [a]. Also, the version 4.0 of LaboImage [8], based on X-Window™ / OSF-Motif™, is designed to be portable on various platforms; this latter version has been developed on Sun SPARCstation™ and ported to several UNIX™ workstations (so far, from the Hewlett-Packard®, DEC®, IBM®, Silicon Graphics® families). Its size is reasonable, being of the order of 4MB for the source code and 2.5MB for the executable code.

In section 2, the features of the working environment provided by LaboImage and its organization are surveyed; the evolution from the first to the current version is shown. The large choice of processing families is enumerated; the tools are then described. Section 3 is dedicated to LaboImage users; the evolution of the interface is described through the multiple versions of LaboImage and a mathematical morphology study is given as an example of utilization. The various, evolving implementation aspects are described in section 4, which covers data structures and interaction methods used in LaboImage, as well as source code organization. The following discussion presents the advantages of such a software for teaching and research; a comparison with other types of softwares is also made. Finally, the diverse experiences accumulated along the development of LaboImage are summarized, as well as its possible evolution.

2. Features

2.1. Working Environment

The LaboImage platform is constructed around a master window, which has two essential functions. Firstly, this master window gives access to a hierarchical menu

---

a. LaboImage is provided free of charge for academic users; the source code is available through anonymous ftp at ftp.teleos.com (pub/VISION-LIST-ARCHIVE/SHAREWARE) and peipa.essex.ac.uk (ipa/src/process); please contact the authors to be kept informed of possible changes.

system pointing to each processing method and tool; secondly, it simultaneously records in a text subwindow and into a log-file the trace of user-selected operations. This organization has been kept since the first SunView™-based version (fig. 1a) to the latest X-Window™ / OSF-Motif™-based version (fig. 1c). Despite the apparent similarities between these versions, many improvements have been brought into the software. In order to stop the never-ending extension of the first level of the menu, a new organization of the menu hierarchy reflects the decomposition of image processing into multiple subdomains. This new hierarchy, available since the SunView™ version 3.1 (fig. 1b), allows an easier integration of new applications into LaboImage and a quick and logical access to each method. This access could be even quicker by using predefined keyboard accelerators. The "Quit" and "Help" options were separated from the remainder of the menu. The interface had also been translated into English, to allow a larger, international use of LaboImage.The independence of LaboImage from commercial packages also became complete since version 3.1, allowing its free distribution.

In order to manage data, LaboImage provides the user with numbered memory planes in which iconic data of any type and size can be stored. Unary operations are performed on single planes, while binary ones require input and output planes. The organization of these planes will be addressed in the implementation section.

2.2. Standard Processing

LaboImage currently comprises over one hundred operators. They can be grouped into different families, according to their utility and complexity. The most basic operations include retrieving an image from mass storage, displaying it and saving it after some processing. Visualization of images is however not the only purpose of this software. The following list briefly describes the classes of operations as presently implemented; the more specialized tools are presented in the next paragraph.

- Display: one or eight-bit depth rasters. A threshold at the mean value with or without error distribution (Floyd-Steinberg algorithm) is used for one bit display; linear or logarithmic grey levels and color (RGB) with or without error distribution are used for eight-bit displays. Postscript™ files can also be generated from the images.
- Storage: scalar, complex or color (RGB) images are saved on mass storage in the LaboImage format (see below) or in the Sun® raster format.
- Retrieval: images in the LaboImage format, Sun® raster format and other specific formats can be loaded.
- Geometry: operations such as rotation, translation, transposition, scaling and symmetry.

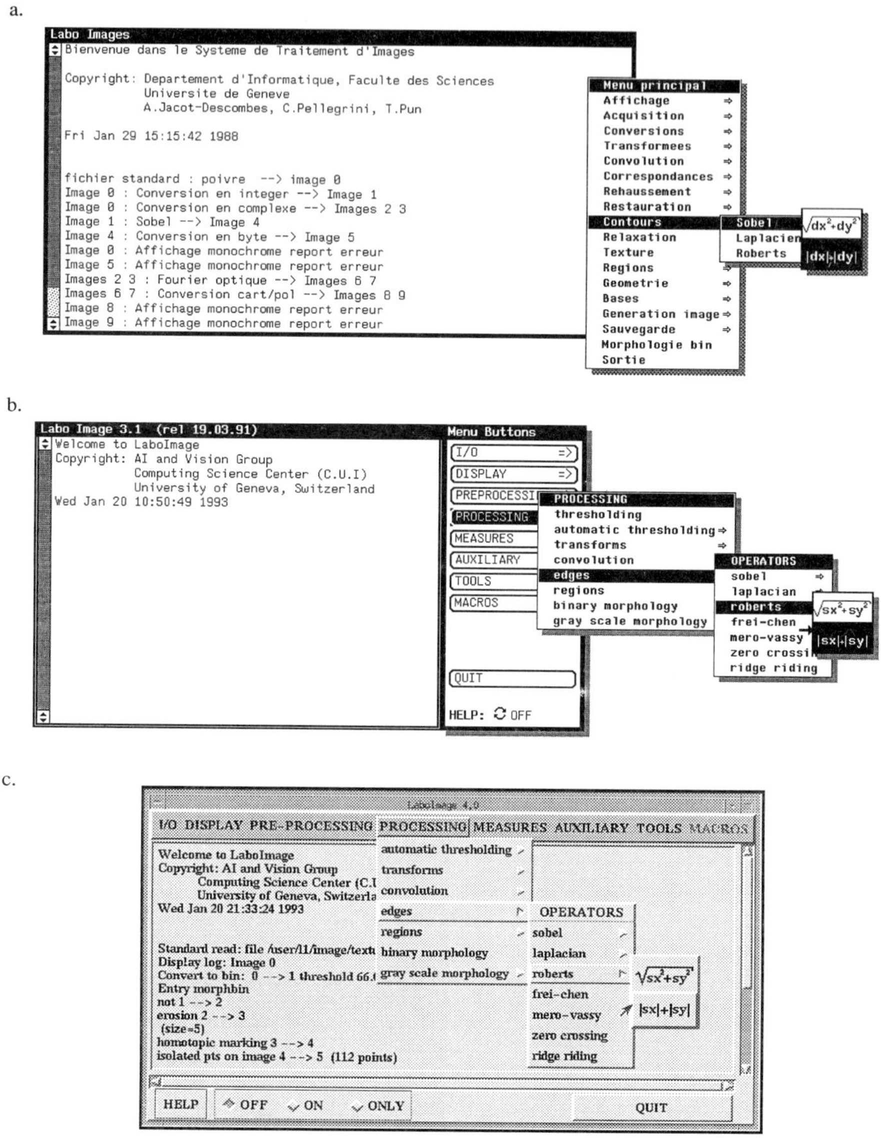

Figure 1: the master window of LaboImage, with its hierarchical menu and operations-log: **a.** the first version (2.0), used with local students (LaboImage 1.0 was only used as a test version); **b.** the first SunView™ version (3.1) available through anonymous ftp; **c.** the latest X-Window™ / OSF-Motif™, public domain version (4.0).

- Histogram: grey level histograms can be computed, displayed and printed.
- Data type conversions: type conversion between the following image types: binary, byte, short, integer, real, cartesian and polar complex. This gives the possibility of modifying grey level scales as well as of trimming the higher and lower parts of the distribution during conversion.
- Arithmetic operations: the four basic arithmetic operations are implemented, with scalars or complex constants and images as arguments.
- Logical operations: NOT, AND, OR, XOR between binary images.
- Measures: some morphological measurements can directly be performed on the displayed images, such as grey level density profiles; statistical measurements can be computed in regions of interest.
- Transformations: regular and optical Fourier transformation, direct and inverse.
- Convolution: users can extract parts of an image or define a mask, to be convolved or correlated with any other image.
- Edge detection: typical operators such as Sobel, Roberts, Laplacian, Frei and Chen, zero crossing and peak following.
- Enhancement: different methods such as histogram equalization, median and high pass filter, Gaussian and Hamming filter, edge preserving smoothing.
- Segmentation: thresholding, regions segmentation methods such as region growing.
- Noise generator: uniform and Gaussian noise can be generated and added to images.
- Binary morphology: all the basic and complex operations of mathematical morphology are implemented in an hexagonal raster [13].
- Grey level morphology: dilation, erosion, opening and closing with different structuring elements or image parts.

The previous list of processing families gives a general view of the software. In fact, many operators allow free choice of parameters, thereby increasing the actual number of available methods. For example, the convolution method permits specification of any mask of any size, allowing for various filter responses. In addition to these processing methods, some basic utilities are included, such as image padding or color to grey level transformations. Table I gives execution times for typical operations.

2.3. Specialized Tools

Part of the specificity of LaboImage lies with the various specialized tools that have been developed for diverse applications, especially in the domain of biomedical

Table I: execution times (in seconds) for typical processing methods.

	SPARCstation 1+		SPARCstation 10	
	256 x 256	512 x 512	256 x 256	512 x 512
histogram equalization	2.1	7.9	0.8	3.1
convolution 3x3	4.7	16.2	2.0	5.0
convolution 9x9	20.7	76.2	6.1	27.2
Fourier transform	9.9	42.5	3.0	10.3

imaging. The list of those tools is given below, with some details for the most important ones.

- Help tool: provides on-line explanations about the operations being performed in a separate, scrollable window. This tool has been developed since the version 3.0; as more and more students and researchers were using LaboImage, while this software was always in development, the on-line help allowed to directly provide users with information regarding new extensions. It also let us avoid having to regularly update and distribute written manuals!
- Interactive color table manipulation: definition of complete look-up tables with RGB or HLS representations [11]. The history of this tool is related to the fact that image synthesis and analysis are taught in one large, integrated course. Students are familiarized with RGB and HLS color representation and look-up table concepts, as well as with the use of pseudo-coloring image visualization.
- Image generation: for interactively designing simple, geometric monochrome and grey level patterns and images of any size.
- Region counting: for regions defined by their contours or by their grey level values.
- Image correction: pixel by pixel correction of an image, making use of a magnifying glass. This for example gives the possibility of closing contours of regions before counting them.
- Expert system: for guiding image segmentation [5].
- 1-D gel autoradiograms analyzing tool [9]: developed for the need of the University Hospital of Geneva. This tool automatically determines the density profiles as well as the profile peaks for one dimensional gels of any type. A graphical interaction allows the user to directly control the process on the displayed gels.
- Image sequence analysis: for comparing images in a sequence. For example, the grey level evolution of a specific region can be measured. This application was developed for a dentistry problem, to quantify the changes of density in lesions of the bone surrounding the teeth in patients suffering periodontitis [3].

- Macro: interactive creation, execution and manipulation of macros. A macro is a succession of processing calls, each line encoding a processing name, the image plane number(s) for input data and/or results and the optional processing parameters.

3. Utilization

3.1. User Interface

LaboImage can run in two modes: interactive or automatic. In the first mode, the software is driven by the menu system attached to the master window, as explained before. It provides users with a hierarchical list of processing procedures and tools, reflecting the image processing subdomains. Users are queried for parameters through specific pop-up dialog windows (called "panel" in the SunView™ terminology and "dialog widget" in the X-Window™ terminology). These windows have also evolved through the different versions of LaboImage. Amongst the improvements is the replacement of a multitude of different, sequential pop-up windows (fig. 2a) by a well designed, unique one (fig. 2b), querying for source and destination plane numbers (i.e. the buffers onto which the processing is made and the result is stored) as well as for parameters specific to the selected method. The systematic use of the keyboard to enter parameters has also been partly replaced by the usage of the mouse, in order to make the interaction quicker and safer.

In the automatic mode, another kind of data (called macro) is used to control the sequence of processings. A macro is a set of operations that are to be applied sequentially. Making a parallel with computer programming, it can be said that macros are akin to programs, in which instructions are calls to processing methods. In order to allow visualization and modification outside LaboImage, macros are stored in ASCII format. In addition, the MacroTool allows the creation and manipulation of such macros. For creation, the user chooses items with the menu system and introduces parameters as in the interactive mode. However, no processing is executed at this stage; the data structure for the current macro is created and updated.

After this creation phase, macros can be run sequentially in three different modes. In automatic mode, the macro is executed as it is recorded, without any changes, i.e. without interaction. This means that each instruction is executed exactly with the same parameters as in the recorded macro. This mode allows for example to load a working state in LaboImage or to run a demonstration program.

In recorded mode, the modification of either the plane numbers or the parameters specific to the method is possible; the pop-up windows are opened as in interactive mode, but parameters are set to those prerecorded in the macro.

In default mode, the specification or modification of processing parameters is possible; the pop-up windows are open exactly as in interactive mode, with the default parameters set to those defined in LaboImage.

The use of such macros has been made available until LaboImage version 3.1. Its major weakness is the lack of control structures, which forces macros to be run a.

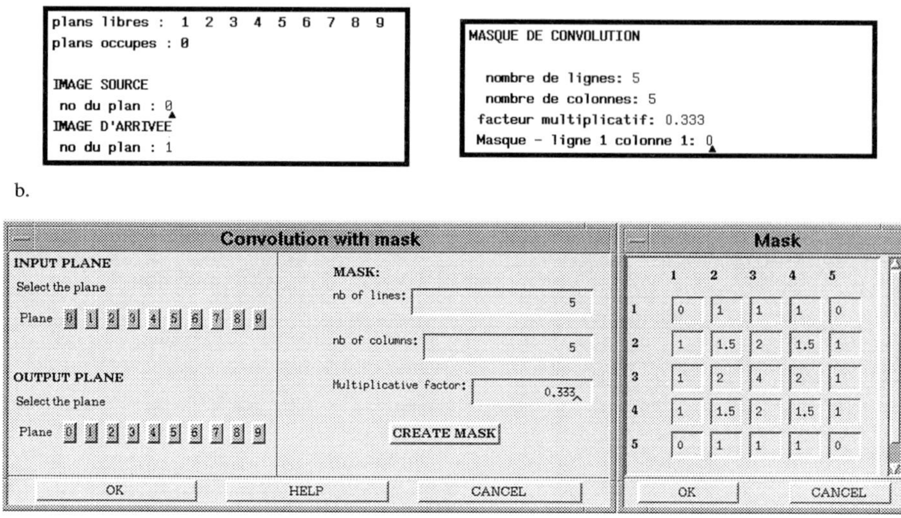

Figure 2: examples of pop-up windows querying for plane numbers and processing method parameters: **a.** the sequence of two windows of the first version, for defining a convolution mask; **b.** the corresponding window of the latest version.

sequentially. This is one of the reasons (except for the lack of manpower!) for postponing the introduction of "MacroTool" in the version 4.0. The definition of a real language to construct more powerful sets of processings and to control their execution sequences is a problem in itself and deserves an extensive reflection.

3.2. Interactive Experimentation

LaboImage has been used since the academic year 1988-1989 as a support to an image analysis course given for computer science students at the University of Geneva. Using such a platform makes it possible to assign exercises that are more interesting and relevant than paper and pencils one. Instead of spending a long time in programming algorithms, students are provided with an interactive toolbox of processing methods. This experimental environment allows them to immediately practice the theoretical concepts seen during the course, thereby implying a better understanding. The proposed exercises can also become realistic image analysis problems, corresponding to practical situations that students may face later. The users can also develop and separately implement their own algorithms, since the simple image file format used by LaboImage makes image exchanges easy. In addition, providing students with a software package guarantees a fruitful harvest of bug reports!

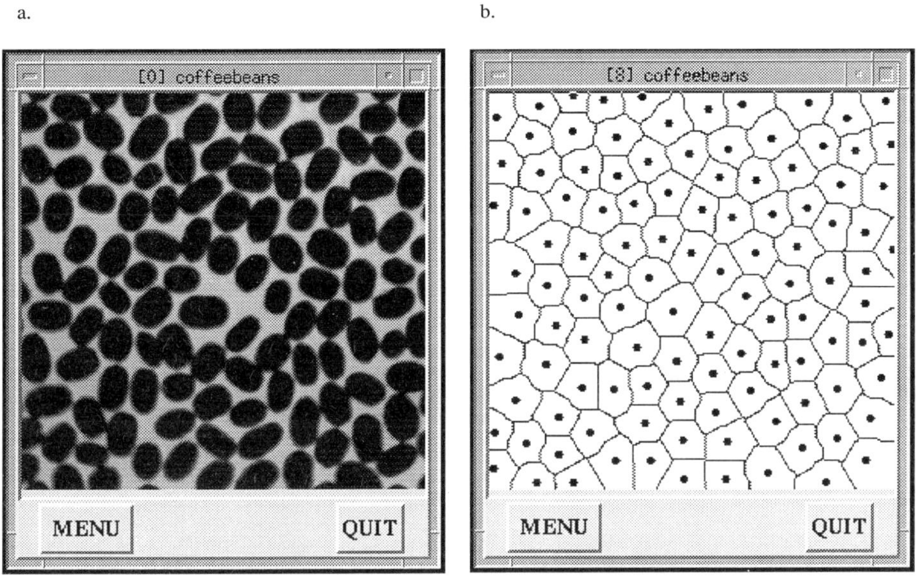

Figure 3: **a.** the original coffee bean image; **b.** each bean has been marked by a dot and the skeleton by zones of influence (similar to the Voronoi diagram) has been computed, using binary mathematical morphology; 112 beans have been detected.

The following example illustrates step by step a typical case of particle counting, that can be solved using binary mathematical morphology. It is performed interactively by choosing all operations with the menus. The first step consists of loading the image to be processed into one of the N available planes; the initial image of coffee beans [2] is then displayed (fig. 3a). After thresholding this image at its mean intensity value, an erosion is applied in order to get a better separation of beans, followed by homotopic marking. Counting the isolated points using a predefined method yields 112 as a result. From the image of isolated-points, the skeleton by zones of influence is then computed and superimposed. The resulting image is finally visualized (fig. 3b). fig. 1c shows the master window containing part of the corresponding trace of operations. If the result is not satisfactory, other threads of operations can be experimented starting at any stage of the processing.

Once a satisfying set of processing procedures is found, a macro can be created by using the same menu system. It allows manipulation of a set of processing steps as a single entity. This macro can later be executed automatically without interaction or just by changing some parameters, as explained in the previous section. fig. 4 shows the ASCII macro corresponding to the set of processing steps used in the current example.

```
I_OLIMSTD TO 0 WITH /user/l1/image/textures/coffeebeans
DISSIMGLO FROM 0
AUXCONBIN FROM 0 TO 1 WITH -1
PROBMO
PROBMOLTRCOM FROM 1 TO 2
PROBMOSTRERO FROM 2 TO 3 WITH 5
PROBMOHLTHMA FROM 3 TO 4
PROBMOHMTIPI FROM 4 TO 5
PROBMOLTHCIPM FROM 5 TO 6 WITH 1
PROBMOHLTSZI FROM 6 TO 7
PROBMOLTRUNI FROM 6 7 TO 8
PROBMODIS FROM 8
PROBMOQUI
```

Figure 4: The ASCII macro for marking coffee beans and computing the skeleton by zones of influence. Each line is a call to a processing method. The name of the method is first coded according to its position in the hierarchical menu, optionally followed by the input (FROM) and/or output (TO) plane numbers and the processing parameters (WITH). For example, PRO stands for processing, BMO for binary morphology, etc.

4. Implementation

4.1. Data Structures

The three major categories of data managed by LaboImage are vectors, images and macros. Each of them is simultaneously available in central memory through a set of N dynamic buffers, numbered from 0 to N-1. The default value for N is currently 10 for each category, but there is no limit to either the number or size of these data structures. Macros are data structures used for controlling the processing methods to be applied, while images and vectors are input or results of processing methods.

Macros are managed in LaboImage through a tool called MacroTool, allowing their creation, execution and manipulation. The external, editable format for macro is ASCII, as in the previous example (fig. 4). Once a macro is loaded into MacroTool, it is compiled; an internal format is dynamically produced, ready to be executed by LaboImage. In such a case, processing methods are selected sequentially in the dynamic structure instead of being chosen in the menu, and processing parameters are managed depending on the execution mode of the macro (see the utilization section).

Vectors (1-D array) and images (2-D array) are data processed or produced by LaboImage. Vectors are mainly used as 1-D work data buffers for color table manipulation or histogramming, while images are input and output of most of the processing methods (table II). The internal format of image and vector data is binary. Externally, the image data is represented by two files, a binary image file (xxx.ima) together with an ASCII descriptor file (xxx.des), as illustrated in table . The original coffee bean image descriptor file is given as an example (fig. 5). This format has not

Table II: *vector* and *image* data structures.

Images	binary, byte, short, integer, float	1 plane
	complex (cartesian or polar)	2 planes
	color (RGB)	3 planes
Vector	color table	1 vector
	intensity histogram	1 vector
	intensity profile	1 vector

Table III: the two files defining an image.

Descriptive file (xxx.des), ascii data		type of image elements(T)
		number of rows (R)
		number of columns (C)
	for each plane	minimum
		maximum
		average
		standard deviation
		date and time of image creation
		comments
Image file (xxx.ima), binary data		R x C elements of type T

```
Type: byte
Row-Count: 256
Column-Count: 256
Maximum[0]: 220.000000
Minimum[0]: 0.000000
Avg-Value[0]: 66.973747
Std-Deviation[0]: 70.832253
Date: Thu Oct 24 15:51:38 1991
Comments: from Brodatz book
```

Figure 5: the descriptor file of the original coffee bean image (fig. 3a).

been modified since the first version of LaboImage, except that the descriptor file format was initially binary. The need to define an image format at the beginning of the project was due to the lack of existing standard. It has been found that representing data by such a couple of files is advantageous; in particular, it allows one to rapidly know the contents of a binary file at hand, and also to easily modify the format of the image description. Using publicly available image format conversion softwares, an image in any format can easily be converted into a binary, raw image file, which in turn can readily be read using the binary image file of LaboImage. The creation of the corresponding descriptor file is done using a tool included in LaboImageProcessing

The current design of processing procedures is again the result of an evolution throughout the multiple versions of LaboImage. The lack of experience in window-based application programming at the beginning of the project and the poor programming structures and tools available to construct such an application were making this initial task rather experimental [6]. The result was a quite complicated model, where processing algorithms, interface building and event handling were not clearly segregated. It was in particular difficult to integrate a new processing method into LaboImage. This model became more and more complicated, as global features such as macros and help were added.

The current version implied many enhancements and modifications in the design of the source code, resulting in a new, clear and well separated model for the integration of the processing methods. Each processing method is now coded as one procedure. The arguments being passed to most of these procedures have a similar structure, i.e. pointer(s) to the input and the output planes, their size, and other possible descriptive information regarding the input plane(s) and the specific processing parameters. This list of arguments is sufficient to warrant independence between the processing procedures and the LaboImage data structures. This feature allows use of these procedures as a library, because neither window-based interaction nor memory access to LaboImage structure is required. It also allows easy addition of new processing procedures to LaboImage. This design is in addition well adapted to a possible transition to object-oriented programming, where an image is an object and processing procedures are methods. In this case, arguments passed to methods should be reduced to the output image and the specific processing parameters, because all input image information will be contained in the calling object.

4.2. Interaction

This section describes the general principle of interaction used in LaboImage. It uses a so-called event-driven scheme, as well for the SunView™ version as for the X-Window™ version. The difference between such an interaction system and a conventional interactive program is presented below.

In a classical program, all components reside in the so-called application part. This means that the control loop, the user interface and the general flow of processing are defined in advance and cannot be altered. Such a scheme is typical of a command interpreter. The software first reads the command; if it is a quit request, the program terminates execution, otherwise parameters are read and the process corresponding to this input is executed (fig. 6a).

In an event-driven application (fig. 6b), the control loop is moved from the application part to the so-called event handler, which reads the events from a stack and calls the adequate procedure according to the received events. This scheme is the underlying structure of LaboImage. When running the program, the master window with its full menu system and associated callback procedures is first constructed in the application part, then control is passed to the event handler. From that moment onwards, the user is just sending events to the event handler by using the mouse and the keyboard. The event handler then analyses the received events and possibly calls the construction

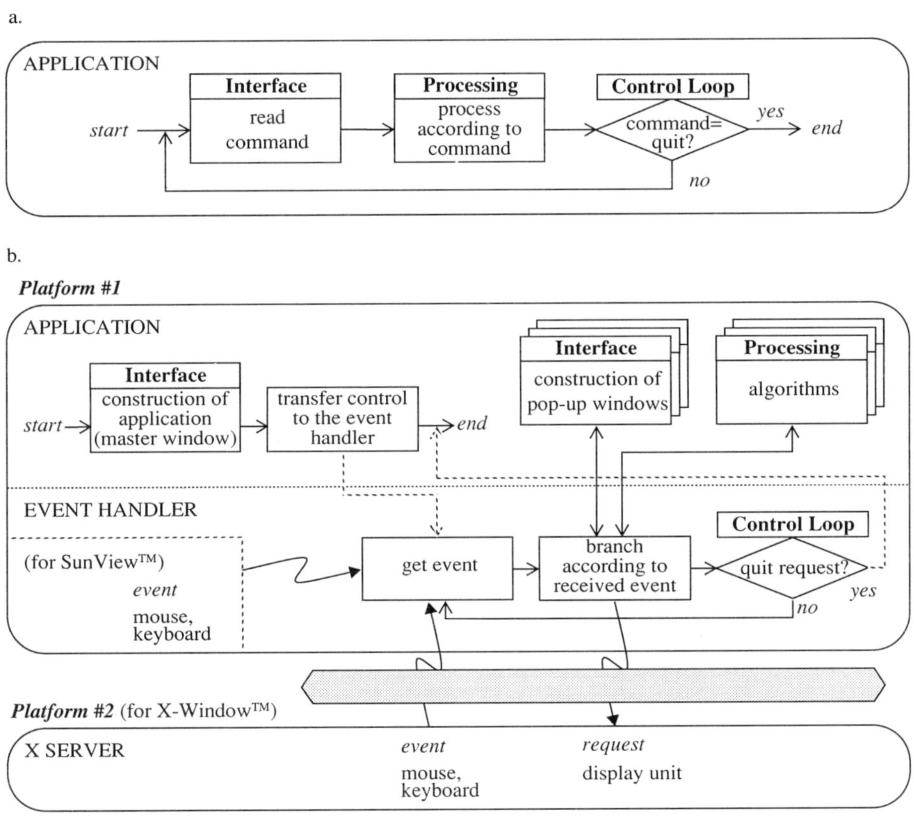

Figure 6: interaction: **a.** conventional interactive program; **b.** event-driven application.

of pop-up dialog windows to ask for parameters. In this case, the event handler receives the parameters from the keyboard and the mouse as events. The appropriate processing procedure is finally called. In opposition to SunView™, X-Window™ is a network-oriented windowing system; another specific feature is thus that the application part and the display part need not be running on the same platform. The communication rules between these two parts through the network is specified by the X protocol.

In summary, a classical program only allows for a very linear flow of operations that is controlled by the program. On the other hand, the flow of execution of an event-driven application is not predetermined, but is controlled by the user. In addition, the X-Window System™ allows one to run applications and display their results on different systems.

4.3. File Organization

The source code organization is very crucial, not only during the development phase, but also for the maintenance and the extension of a system. The LaboImage source code files have been carefully organized, reflecting its event-driven nature as well as the decomposition of image processing into multiple subdomains. The current organization is schematized in fig. 7. In order to avoid a proliferation of files and the subsequent naming problem, processing methods have been grouped into entities reflecting the subdomains of image processing. This grouping is identical to the one reflected by the menu structure. For example, the Laplacian method belongs to the edge detection methods that are callable from the edge submenu, and therefore has been put into the edge detection set of source files.

Two remarks can be made regarding fig. 7. First, LaboImage is made of a core consisting of two files (*LaboImage.layout.c* and *LaboImage.callback.c*), and of a set of

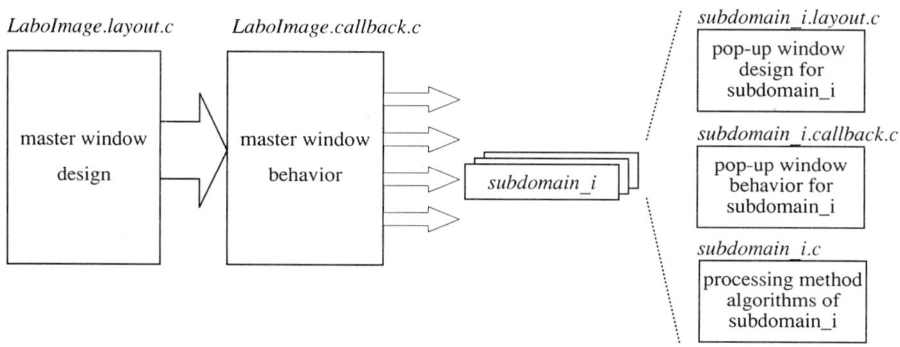

Figure 7: organization of LaboImage source code files, reflecting the same decomposition of image processing into subdomains as in the hierarchical menu system.

subdomains, each one consisting of three files (*subdomain.layout.c*, *subdomain.callback.c* and *subdomain.c*). Second, the core as well as the subdomains contain file names with the suffixes *layout* and *callback*. The interface design is constructed by the layout files, whereas the behavior of the interface is coded into the callback files. *LaboImage.layout.c* is therefore devoted to the construction of the master window, consisting principally of the hierarchical menu system, while *LaboImage.callback.c* describes the behavior of each item of the hierarchical menu, i.e. which procedure to call once an item is selected. In each subdomain, the layout file is devoted to the construction of the pop-up dialog windows specific to each processing method, whereas the callback file describes their behavior. Once the processing parameters have been validated, the processing procedure residing in the third file of a subdomain is finally called.

5. Discussion

It is interesting to situate LaboImage in the realm of imaging softwares. Several aspects can be considered, such as complexity, interface or scope. Our initial aim was to develop a user-friendly environment for supporting an image processing and analysis course. At the time of the preliminary study (1987), and to our knowledge, no such software was available. Quite powerful libraries of routines (such as Spider) were in favor, but no interface, even a simple terminal one, was existing to facilitate their use. The motivation of our project was therefore the development of an experimental, window-based environment giving a easy access to standard methods; this situates our approach at the opposite of softwares such as the PIC System II [16], which provides elaborate algorithms through a simple terminal interface. In the same way, no standard image format was emerging at this time; this obliged us to define our own format. A simple format with the image and its description standing in two separate files has proved to be a good choice, because it is now easy to translate data to or from other image formats.

Some imaging softwares have been designed for more specific tasks; one example is KBVision [15], which provides an elaborate environment for the prototyping of vision oriented knowledge based systems, and consists of elaborate data structures and visualization tools. Other softwares have more general objectives, such as Khoros [12] which offers a development environment for data processing, that uses the visual programming concept. It seems that each of these softwares is developed according to a strategic goal. After about 15 years of research, KBVision aims at a commercial success, whereas the Khoros consortium wishes to suggest a standard environment for data processing. LaboImage is rather an academic freeware, available for the academic community, for teaching as well as for research. Its simple structure makes the integration of other applications quite easy, such as work done for master theses, or applications developed for a specific task. These collaborations are beneficial for LaboImage users, which have access to an increasing numbers of features from the software.

One strong feature of LaboImage lies precisely in this integration of standard processing methods (such as filters, transforms, etc.) together with particular tools. As a typical application, it is possible to take an input image of a 1-D gel, filter it to remove acquisition noise, then perform a 1-D gel analysis [9]. The results are kept within LaboImage, allowing for example comparison of several processed gels by arithmetic subtraction. Another example of a specialized application in LaboImage is the analysis of superimposable dental radiographs [3], to study the evolution of the jaw bone density in patients treated for periodontitis. LaboImage is therefore evolving according to specific needs; new tools are regularly being added. It is also planned to use the Osiris [10], Melanie [1] and LaboImage [7] imaging softwares for regular practice at the University Hospital of Geneva.

A significant amount of experience has been accumulated in developing such a software. The initial utilization of the local language (French) for its interface rapidly became an hindrance to its diffusion. This choice was motivated by the fact that LaboImage was first devoted to our students. However, English is so widespread in the computer community that the English version has now been adopted for teaching. The development of a shareware must also be independent from commercial products in order to be distributed without restriction. Although the use of routines from the Spider library was initially a saving of time, allowing us to concentrate on the interface development, it deprived potential LaboImage users of the most juicy functions. This lead us to rewrite every function, a good student assignment by the way. The same attention has also been paid for the choice of an interface builder for the development of the X-Window™ version. Quite a few such products need a costly run-time library for the developed application. The only non-public domain dependence we accept is the OSF/Motif™ windowing system, because it seems to become a widespread standard. In addition, the ownership of libraries is not required in order to run a statically compiled version of LaboImage under other X windowing systems.

Finally, developing such an academic software represents an important time investment, especially since our research group is rather involved in computer vision than in image processing. It could well be this lack of available manpower for developing and maintaining LaboImage that made it a public domain instead of a commercial software!

6. Conclusion

Digital image processing and analysis is very useful for processing various types of scientific data, but requires powerful and user-friendly softwares for non-specialized users. This theoretical branch needs also to be put into practice in order to be mastered. LaboImage is an answer to such a need. It provides researchers and students with various image processing methods through an easy-to-use interface. The major advantage of such a software is its large variety of processing procedures and specialized tools: many different problems can be solved using a unique system. It also provides the possibility of comparing various methods and of designing well adapted solutions.

Finally, it is emphasized that the implementation model clearly separates the interface part of the system from the processing one. This conception allows easy development, maintenance and extension of LaboImage to new theoretical methods or practical tools.

7. Acknowledgments

We gratefully acknowledge the work done by all researchers and students which have been or still are actively involved in this project. We would like also to particularly thank Dr. D. F. Hochstrasser and Dr. O. Ratib, from the Medical Imaging Group of Geneva University Hospital, for their extended support, as well as M. Logean-Rupp for her intensive work; LaboImage 4.0 could not have been without their help. We also wish to thank Prof. C. Pellegrini, head of the Department of Computer Science, for his constant help during those years.

8. References

1. R. D. Appel, D. F. Hochstrasser, M. Funk, J. R. Vargas, C. Pellegrini, A. F. Muller and J. R. Scherrer, "The Melanie Project: From a Biopsy to Automatic Protein Map Interpretation by Computer", *Electrophoresis*, vol. 12, 1991, pp. 722-735.
2. P. Brodatz, *Textures: A photographic album for artists and designers*, Dover, New York, 1966.
3. B. Dubrez, A. Jacot-Descombes, T. Pun and G. Cimasoni, "Photodensitometric vs. high-resolution digital analysis of superimposable dental radiographs", *Journal of Dento-maxilo-facial Radiology*, vol. 21, February 1992, pp 40-44.
4. R. C. Gonzalez and P. Wintz, *Digital Image Processing*, Addison-Wesley Publishing Company, Reading, Mass., 1987.
5. Z. P. Hu, T. Pun and C. Pellegrini, "An Expert System for Guiding Image Segmentation", *Computerized Medical Imaging and Graphics*, vol. 14 (1), 1990, pp. 13-24.
6. A. Jacot-Descombes, *Laboratoire Image: développement d'un environnement convivial pour le traitement interactif et automatique d'images*, Master thesis, University of Geneva, Switzerland, 1988.
7. A. Jacot-Descombes, M. Rupp and T. Pun, "LaboImage 4.0: Portable window-based environment for research in image processing and analysis", *MedInfo 92, 7th World Congress on Medical Informatics*, Geneva, Switzerland, 6-10 September, 1992.

8. A. Jacot-Descombes, M. Rupp and T. Pun, "LaboImage: A portable window-based environment for research in image processing and analysis", *SPIE Symposium on Electronic Imaging Science and Technology, Image Processing: Implementation and Systems*, San Jose, California, USA, Feb. 9-14, 1992.
9. A. Jacot-Descombes, K. Todorov, D. F. Hochstrasser, C. Pellegrini and T. Pun, "LaboImage: a workstation environment for research in image processing and analysis", *Computer Applications in the Biosciences*, vol. 7 (2), 1991, pp. 225-232.
10. Y. Ligier, M. Funk, O. Ratib, R. Perrier and C. Girard, "The OSIRIS User Interface for Manipulating Medical Images", in *Proc of NATO ASI meeting on "Picture Archiving and Communication System (PACS) in Medicine"*, Evian, France, Oct 1990.
11. R. W. Lutz, T. Pun and C. Pellegrini, "Colour Displays and Look-up Tables: Real Time Modification of Digital Images", *Computerized Medical Imaging and Graphics*, vol. 15 (2), 1991, pp. 73-84.
12. J. R. Rasure and C. S. Williams, "An Integrated Visual Language and Software Development Environment", *Journal of Visual Languages and Computing*, vol. 2, 1991, pp. 217-246.
13. J. Serra, *Mathematical Morphology and Image Analysis*, Academic Press, New York, 1982.
14. *SPIDER User's Manual*, Joint System Development Corporation, Tokyo, 1983.
15. *The KBVision™ System*, release 2.5, Amerinex Artificial Intelligence, Inc., Amherst, MA, May 1992.
16. B. L. Trus, M. Unser, T. Pun and A. C. Steven, "Digital image processing of electron micrographs: the PIC System II", *Journal of Scanning Microscopy*, accepted for publication, 1993.

SciIlmage: A Multi-layered Environment for Use and Development of Image Processing Software

Richard van Balen   Dennis Koelma   Ton K. ten Kate   Benno Mosterd
Arnold W.M. Smeulders

Faculty of Mathematics and Computer Science, University of Amsterdam
Kruislaan 403, 1098 SJ Amsterdam, The Netherlands
tel.: +31 20 525 7516, e-mail: ricci@fwi.uva.nl

August 23, 1993

Abstract

This paper presents an integrated system made up of a portable multi-layered interactive environment for software development and use, and a class based infrastructure for image processing.

The interactive environment is responsive to different type of users, and consists of a library handler, a C interpreter, a command expander, a menu and dialog generator, and a visual programming interface. All interface layers are generated from a command description file. The design considerations of each of the interface layers is discussed.

The image infrastructure enforces strict image class separation, and takes care of actions common to all image classes. Through the use of an abstraction mechanism based on function overloading, the infrastructure allows image sub-classing and the definition of new image classes, which automatically inherit the consistent behavior of the total environment.

An overview of the extensive image processing libraries functionality is given, with special attention to bit-mapped binary images and mathematical morphology advancements provided in the system.

1 Introduction

In image processing, interactive environments have always played an important role. Generally there are no off the shelf solutions available for image processing problems. Although the theory of image processing has advanced well beyond development based only on trial and error, the amount and variability of data, even within similar images, is so large that visual evaluation of attempted strategies is a practical necessity [1]. The immediate feedback interactive systems provide reduces the time needed to develop new applications. Furthermore, the ability to see the result of processing and to modify code or parameters within seconds brings new insights, such as how sensitive an algorithm is to a small change in the image, or how parameters should be tuned in response to certain shapes found [2]. Obviously, an interactive system is of great importance to image processing design, methodology and development.

Unlike word processing, for example, the processing and interpretation of digital images still is in many respects an evolving discipline. In 2D and especially in 3D image processing many issues remain unsolved. The general situation is that there is an ongoing flow of new or renewed algorithms and implementations. As a consequence software environments for image processing are still expanding and evolving [3, 4].

Since dedicated hardware systems are targeted at speeding up stable methodology, using them is in contrast with the state of the art in the field requiring to quickly add new algorithms [5]. The natural alliance between interactive systems and image processing, combined with the need for flexibility, has led many research institutions to develop their own image processing environment. Today close to a hundred image processing systems are in existence [6] prohibiting the exchange of algorithms between research groups [3, 7]. The need to exchange software and ideas is especially important in the study of new research topics, such as model-based image processing and image understanding, since the complexity of these topics force research centers to cooperate [3].

In a survey covering 72 image processing languages, Preston concluded that the diversity in image processing environments is not justified since the image processing applications themselves are not so disparate as to require unique and specialized data types or different programming constructions [6]. It is therefore recognized that, if an image processing environment is to meet the requirements of multiple research institutions, the time is felt right to specify a portable and flexible environment to allow for ongoing changes in software methodology [3, 5].

In a research context there is often no clear distinction between an image processing user and the developer of the program [8]. Computer scientists typically develop or at least adapt programs and algorithms for their own use. In the process of developing new algorithms they frequently switch between a programming mode and a users mode. As a consequence most image processing systems provide an algorithmic view on image processing. The user is bothered with explicit image type and dimension management, and restrictions on image input and output designation. Furthermore the user is required to know of algorithmic details, such as whether an operation can be performed in place or not. To serve programming and use of an image processing system beyond a research setting it is in our view, therefore, essential to provide a logical and dynamic view on image processing rather than a static and algorithmic one.

The various image processing environments range from a collection of routines in software libraries [9], via command [10], menu [11], or visual programming [12] driven systems that are easy to use, to highly specialized but idiosyncratic programming languages [1, 13]. Although the latter group contains variable types and programming constructs, one is often confronted during application development with limitations caused by the specialized nature of the language. The importance of programmability is stressed in [1, 2, 14]. In [15] Graham states that in the design of an image processing system it is all too often the case that little attention is paid to general programmability. We share the opinion that attention to this topic pays dividend, but we stress the importance of adhering to existing programming standards rather than defining a new one.

Different forms of environments each provide a suitable interface for one group of users in particular. Different type of users, however, have different interface needs [16]. Menu based systems, for example, are easy to learn for novices, but they are a nuisance for expert users as they are forced to travel through (multiple) menus to specify an action. A visual programming interface can be so intuitive that even users with no programming experience can create simple, but non-trivial programs. Experienced program developers, however, need complete control and program constructs which are very difficult to express with visual programming.

Also the users themselves are not a constant factor since they progress from beginners, to experienced, to possibly experts, and may eventually need special features not presented in the environment, forcing them to program the missing aspects themselves. We therefore conclude that the interface of an interactive environment should be such that it is responsive to different types of users. Moreover, if an environment simultaneously provides an interface that accommodates both experienced developers as well as lay end-users the barrier between development and use of a system is resolved.

Requirements

From the observations made above we produced the following list of requirements for a new and flexible environment for image processing.

- The environment should combine the advantages of different systems, being helpful to beginning users yet providing more power and flexibility when the user gains more experience.
- To also allow non-programmers to construct simple programs a visual programming paradigm should be included.
- Since no environment for image processing can possibly supply the answer to all the users needs and questions, the environment should not only allow the use of image processing software, but also allow for software development.
- Ideally further development within or with the environment should take place in a standard programming language.
- The environment should not be limited to the image processing domain. This would ideally mean that the environment also allows easy integration of different application domains.
- The consequence of the previous requirement is to clearly separate the environment from the image processing algorithms. A flexible image processing infrastructure should, therefore, be designed independent of the interactive environment.
- Both the interactive environment and the image processing infrastructure should be portable for a variety of machines and operating systems.

In this paper we address the computer science aspects of a general purpose interactive environment [17] and a flexible infrastructure for image processing [18]. The two combined results in a multi-layered environment which meets the demands of portability, maintainability, and flexibility necessary to function as a general vehicle for use and development of image processing software.

Section 2 features the interactive environment and a description of each of the interface layers. Section 3 elaborates on the object oriented image infrastructure. In section 4 an overview of the image processing functionality is given and some important aspects of the image libraries are discussed. Section 5 is on related work and use of the system in several projects. In section 6 a discussion and some conclusions on the presented environment are given.

2 SCIL: A Multi-layered Environment

ScilImage is the combination of the user front-end environment SCIL [17] and image processing libraries under the name of Image [18]. SCIL (Standard C Interpretive Language), can be tailored to many different application domains, and consists of multiple interface layers, which all are active at the same time. Figure 1 shows an overview. Compiled functions are interfaced through the library handler, which is practically invisible to the user. The first user visible layer consists of a C-interpreter. In the second layer a command expander working on image processing commands

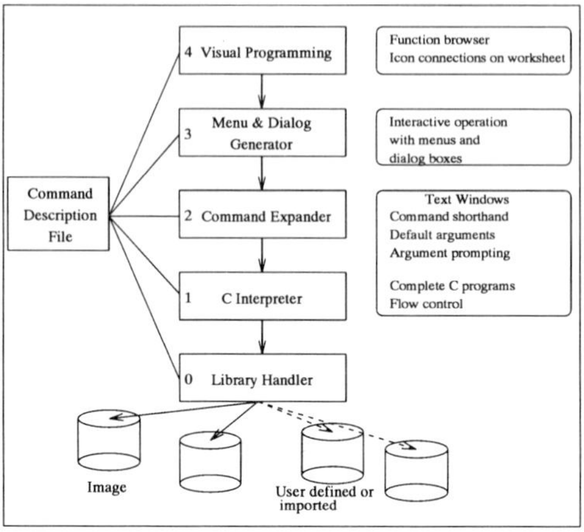

Figure 1: The SCIL interface layers. All layers are generated from the Command Description File.

known to the system opens the way to efficient interactive (image processing) design, hiding the C-level from the novice and allowing shorthand typing for the experienced user. In the third layer, a window management system generates menus and dialogs for command selection, now hiding the command level from the naive user.

The layers in the front-end offer different levels of user support. Going from the C interpreter, via the Command Expander level, to the Menu and Dialog generator, the system offers increasingly more user-friendliness but looses on flexibility and speed of interaction. At the menu and dialog level, for example, no program flow-control is available. This is reserved for the C interpreter level. There is one more interface layer supporting visual programming to combine user-friendliness with programmability. It is based on a data-flow paradigm, which allows users without any programming experience to built simply structured but non-trivial programs [19].

All interface layers are generated from a Command Description File (CDF) which insulates the interface layers from the application. We discuss the design considerations of each of the interface layers and go over the functionality of each of the layers one by one.

The C interpreter

We have selected a standard programming language as the basis for command specification. The advantage is that the command language is well defined, generally known, and equal to the language used to run actual programs. As a consequence interactively developed programs can also be used in compiled mode. C is the obvious choice as most software and hardware based image processing routines are interfaced through C-callable libraries. The C interpreter ensures general programmability and portability of the developed applications. The use of the interpreter not only shortcuts the traditional edit-compile-debug cycle, but also allows the developed applications to be compiled and executed as stand-alone programs.

The Library Handler

Linking of external libraries takes place through the library handler, a separate layer of the environment which is invisible to the user. It provides the ability to add new commands quickly and easily. When adding functions to the environment the user only deals with a command description file, adding one entry for each function to become known to all the layers in the environment. There is no need to make any changes to external libraries. In this way the separation between the environment and application code remains clean and well defined.

In practice, there is one command description file per library. All command description files are concatenated into a single command description file, from which a table with the names, addresses and the number of arguments is compiled, and linked in with the rest of the system. The major advantage of this approach is that software libraries of various engineering fields, such as image processing, data-basing, statistical analysis and graphics, can easily be combined in the system. It also allows for easy software exchange between different users, as they can exchange personal libraries by sending the library in compiled form together with its command description file.

Rather than tailoring the C-interpreter itself to the image processing domain, the ability to add any software library to the C-interpreter, broadens the usability and scope significantly.

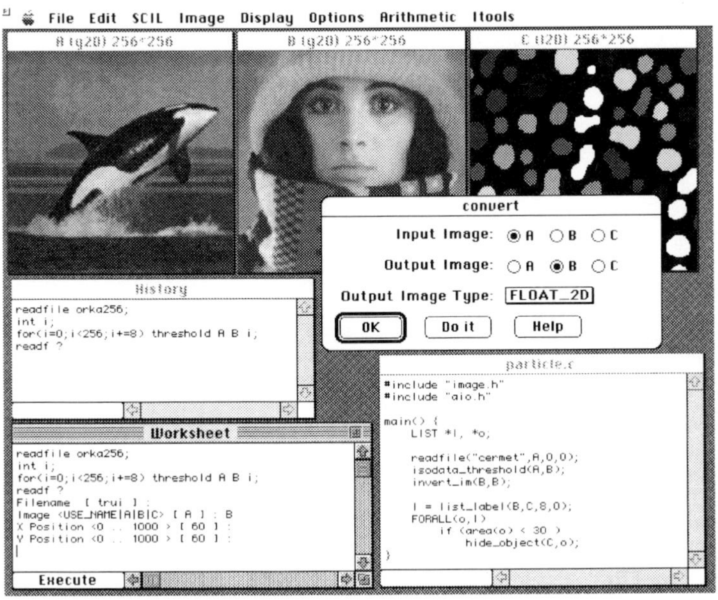

Figure 2: Impression of the Macintosh version of ScilImage with several interface layers: A generated menu bar, dialog box, three images, a run-editor containing a C program, a history window, and a worksheet showing C statements and commands.

The Command Expander

Where the benefits of a C interpreter are obvious to program developers, they were found too complex for non-programming users. Non-programming users have difficulty or find it cumbersome to type all C punctuation. To solve the issue, the command expander responsible for transposing (possibly abbreviated) commands into legal C function calls, serves as an extra interface layer on top of the C interpreter in layer 2 of the total environment.

At the command expansion level arguments are checked against legal values. When arguments of a command are left unspecified the list is completed with default values. Interactive argument prompting is invoked when requested by typing a question mark, when an illegal value is encountered, or when automatic conversion to the correct type fails.

An example of the command expander functionality can be seen in fig 2. As all other interface layers, the command expander operates on the command description file.

The Command Description File

Per processing function the command description file contains a specification of each argument including the type, default and range or legal values, and explanation of the prompted argument. Apart from the standard C types, such as "int", "float", or "long", some special types are supported, e.g. "odd", "even", "choice", "infile", "outfile". A typical function entry in the command description file is given below:

```
FUNC    uniform
MENU    Filter
ICON    uniform.bitmap
ARGS
        image   in    A    -    -    Input Image
        image   out   B    -    -    Output Image
        odd     con   3    1    61   X Filter Size
        odd     con   3    1    61   Y Filter Size
#
```

The "choice" type allows the user to specify a single choice out of a set of legal choices. It plays an important role in the command expander. As a detail, consider that through the use of names with associated aliases it connects symbolic names to constants, particularly useful in image processing development to avoid the introduction of "magic" numbers.

In the system, new types can be defined through function hooks, that is calling a specified function if an unknown type is encountered. An example of such a foreign type is the "image" type. If an image type is encountered in the command description file, a hooked function is called, transposing the current images into a legal choice type consisting of the image names (with the image address as alias), which the command expander can handle.

The Menu & Dialog Generator

The Menu & Dialog (M&D) Generator in layer three of the environment dynamically creates at run-time a window and menu based interface from the specifications in the command description file. Mouse activation automatically shows the available commands in cascading pop up menus. When a command is selected a dialog box is created giving an overview of all the arguments of the function together with their defaults and ranges or legal choices whichever is relevant. Depending

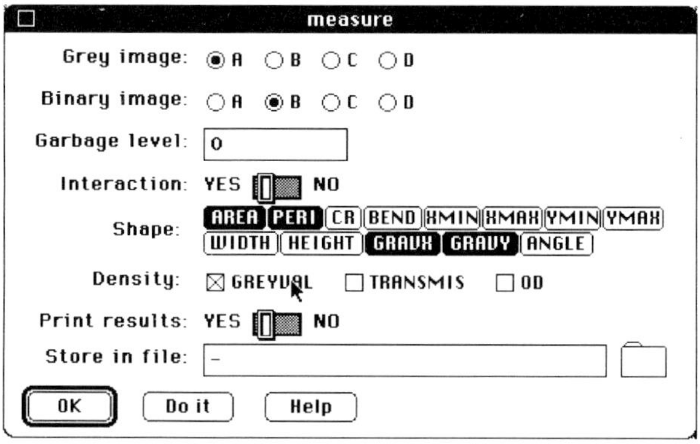

Figure 3: A dialog box as automatically generated from the Command Description File

on whether a help file can be found for the command at hand also a help button is created. Fig 3 shows an example of a generated dialog box.

The M&D generator performs an educated guess, based on the argument type and legal values, to choose the most appropriate graphical attribute to display in the dialog box, such as slider, cycle, sequence of buttons, or editable text field. In this way additional specifications in the command description file are not needed, serving simplicity in use, consistency throughout the user interface and dynamical adaptation of the attribute.

To illustrate the dynamic adaptation in more detail, consider the subsequent creation of a series of images. When there are only a few images, each image is symbolized by a button with the image name next to it. When some new images are created, the name of the images is placed inside the buttons to save space. Finally, if the number of images exceeds a second limit, say ten, only the current selected image is shown allowing the user to choose another image from a menu, or by cycling through the available images.

Since the interface is generated at run-time, the listing and sequence of commands in menus is tailored quickly to personal preferences or customized for a particular application.

The Visual Programming Layer

The SCIL environment also has a visual programming (VP) layer based on a data-flow paradigm. The rationale behind the use of visual programming is discussed in [19, 17]. It is obvious that VP is user-friendly, and visual programs offer a quick insight in the program structure. Especially for image processing they are helpful as the transformation of the image data can be followed and visualized through interacting on the program flow.

For an overview of different visual programming paradigms and systems the reader is referred to [20, 21, 22].

We have opted, apart from having visual expressions, to also implement a C-icon which can hold any C-statement, allowing a mixture between visual and textual expressions, to circumvent the problem that not every programming issue can easily be expressed in a visual manner.

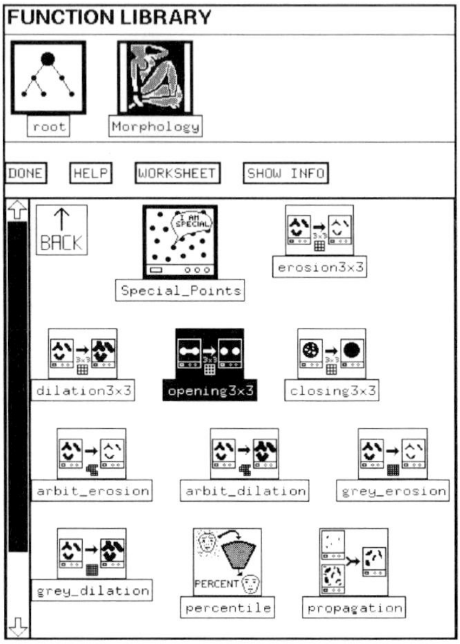

Figure 4: Part of the VP layer: the function browser

The VP layer consists of two parts; a function browser and a worksheet. The function browser (see figure 4) is similar in structure to the main menu, since both are generated from the command description file, only now the different menus and menu items are represented as named class-icons and function icons.

When a function icon is selected it can be placed into the worksheet. The worksheet (see figure 5) consists of two parts, a workspace in which the visual program is built and a control area with command buttons and edit icons.

In the workspace the function icons can be connected by selection of input/output pin pairs. As soon as two pins have been selected, and the pin types match, the connection becomes visible. There can only be one connection to an input pin, but an output pin can be connected to multiple input pins.

Through a number of edit icons the user can manipulate the visual program in several ways. We only mention a few of the edit commands here: the user can obtain help information, destroy components, execute selected functions or the complete program, compose a part of a visual program into a single icon, start a separate worksheet to edit a composed icon, or add a C icon to mix C statements into the visual program. When a visual program has been developed and is working

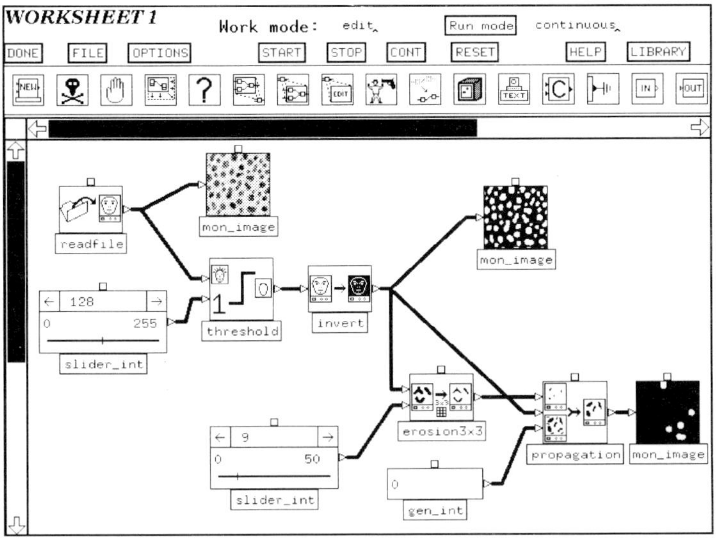

Figure 5: Example of a worksheet

correctly it can also be saved as an ordinary C program. This allows the user to execute the program without the visual programming environment.

Visual Program Execution

Once the program has been composed it can be executed (in a continuous or step mode) by pressing the start button. Since the program is interpreted execution starts immediately. The order in which the functions are executed is a combination of data and demand driven. If there is a function that has valid data on all input pins then it is executed (data driven). If a function does not have valid data on all input pins its predecessors are executed (demand driven). The function under execution is highlighted so the user can experience the execution.

Like all other layers the VP layer operates on the command description file. For the VP layer each function argument in the command description file can be described as being an input, an output, an input-output, or a control parameter. Input and output parameters appear as pins to the left and right of the function icon respectively, while an input-output parameter flows through the icon. Control parameters are parameters that seldomly need be changed. They are not represented as pins. If a function icon has control parameters a small rectangular handle is placed above the icon, which, if selected, allows the user to change the control parameters through a generated dialog box.

3 A Data-types Infrastructure for Image Processing

From a software engineering point of view, a number of observations can be made when studying the nature of images and image processing.

Image Classification

Although all digital images basically consist of arrays of pixels, there is a natural basis for the division of images into different classes. The difference between grey valued images and binary images is clear. Less obvious is the difference between a grey valued image and a label image identifying connected components, for grey valued images and label images use the same fields to store the data. Still they are essentially different classes of images as they give a different semantic to the pixel values stored. As a consequence grey valued and label images have different sets of admissible operations. This natural way of classifying images by their admissible operations should be reflected in the design of an image processing system. An image processing system must, therefore, enforce *class separation* of its data: the digital images.

Most existing image processing systems lack strong image class separation. An example of weak class separation is the representation of binary images as bit-planes in a grey value image. This representation not only spoils memory and transport speed, but also prohibits *class safety* since the content of a binary image can change as a side effect of an operation on the grey value image.

Apart from class safety, another argument for strong class separation is execution efficiency. We cite Piper and Rutovitsch [23] that for efficiency reasons an image processing system should not impose any data representation on images, but allow images to be represented in the most suitable way for processing.

Object Oriented Infrastructure

Class separation calls for an *object oriented* approach in image processing systems [24]. Since it is sensible to deal with actions and properties common to all image classes at a generic level, while dealing with more specific actions in specialized classes, we argue that an *object oriented infrastructure* should serve as the backbone in a growing and evolving system.

To allow the infrastructure to detect whether an action is legal on a specific image class, a list of admissible actions is known to each of the image classes. Furthermore class safety, consistency, and standardized checks on function arguments, are taken care of by the infrastructure, rather than being the concern of every programmer.

The infrastructure provides the possibility to create new image classes. Through an abstraction mechanism new classes automatically inherit the consistent behavior common to all image classes. In this way the evolving nature of an image processing system can be kept consistent with respect to the data-structures. It anticipates to the need of future image classes without having to change the infrastructure. Future image classes include image sequences, time series, and multi-scale images.

Abstraction Through Function Overloading

Although images are strictly separated by class it would be undesirable and uncomfortable to an end-user if an operation would carry a different name for each of the image classes. To map a single action name to different image class specific functions, abstraction is handled through function overloading. Function overloading introduces a specific function formalism, and saves considerable on the number of function names visible to the user.

Function overloading also opens the way to action sharing and inheritance, thus allowing image classing and sub-classing to be used as an organizational tool. For example, it is possible to design a set of operations targeted at specific processing domains such as microscopical images or the domain of radiology images. Sub-classing a new image class from the grey value main class, reflects the specialized nature. It organizes the set of operations such that it helps both the user and the program developer by restricting the number of admissible actions.

3.1 The infrastructure: Services and Behavior

The infrastructure communicates with different image classes in an abstract way through a table driven function overload mechanism. By making a limited number of actions obligatory for every image class the infrastructure generalizes common actions. Thus it standardizes overall system behavior, regardless of any present or future image class.

Dynamic and Logical Images

In the system, the user views of an image as a named object capable of holding any image class, and lets the program concern about dimension management, class safety and legality of actions. It implies that the user is not restricted in choosing what logical image is going to be the output image. As a consequence, an image object is omnipotent in the sense that it can change its class and dimension. So an image object must serve as input as well as output to an operation if so desired, even if the image class of the object changes as a result of the operation. The infrastructure automatically takes care of the adjustment of the output image in all situations by overloading the obligatory *create_image*, and *destroy_image* actions, shielding the user from implementational details.

Generic Region of Interest Processing

Especially when dealing with very large images, there is a need to process only an arbitrary part of the image: a Region Of Interest (ROI). ROI processing is considered a concept at the general level, both in the sense that it is independent of any image class, and in the sense that it should apply to the complete range of possible image processing actions. As a consequence, ROI processing belongs at the level of the general infrastructure, rather than at the algorithmic level.

Generic handling of ROI processing, requires the possibility to copy a (hyper) rectangular part of an image forming a temporary image on which the operation can be performed. After the operation the data is copied back to the original image. By limiting the pixel access by a bit-mask, arbitrary shapes of ROI's are handled. Through overloading the *copy_part_image*, and *copy_masked_part* actions the infrastructure deals with ROI's independent of image type. With this approach, the processing algorithm has no concern with ROI's since a ROI in itself is just an image.

Image Class Conversion

In a class based system as described here, there is a need for class conversions. A general solution to the class to class conversion has been achieved by letting every image class store a single row in a common format. Row by row the class specific data format is converted to a common format and on to the new format. Each image class is expected to overload the *convert_to_common* and *convert_from_common* actions. Once implemented, these actions are invisible to the end-user who only deals with the abstract function *convert_image*.

Abstraction and Automation of Image Display

In general, the display of an image not only depends on the image class but also on the image processing domain. In radiology or microscopy, for example, visualization takes on an essential role, whereas in a theoretical research environment the requirements are often much simpler, or different anyway. However, since function over-ruling is implicit to function overloading, different display requirements are solved generically.

The infrastructure provides automatic image display. To also enable the display of a ROI, overloading of an action to display an image part is obligatory for every image class.

4 The Image Libraries

ScilImage offers an extensive set of image processing operations, separated by image class into different libraries. Table 1 shows the presently supported image classes of the standard.

Binary 2D	Binary 3D
Grey integer 2D	Grey integer 3D
Label 2D	Label 3D
Grey float 2D	Grey float 3D
Complex 2D	Complex 3D
Color 2D	Color 3D

Table 1: Image classes presently supported by ScilImage standard

All class specific libraries are based on the image infrastructure. A *'generic'* library, containing all user visible function names, serves as an abstract placeholder to reach class specific functions through function overloading. To interface the image libraries to the SCIL environment an entry for each of the functions in the 'generic' library has been made in the SCIL command description file. The overhead imposed by SCIL or the overloading mechanism is neglectable, as the time needed to parse and call a function is only a fraction of the time needed to execute the image processing operation.

In ScilImage, attention has been given to image processing implementations which are tailored to the specifics of general purpose computers [11]. The obvious advantage of these fast software implementations over specialized hardware is that they can easily be ported to other architectures e.g. to benefit from faster execution, or to be integrated in other systems and to become more widespread in use.

At present more than 300 image processing operations are visible to the user. The actual number of different functions is much larger since many of these operations are overloaded to work on several image classes. Moreover, it should be emphasized that the number of functions does not say all about the power of a system. Sometimes, the functionality of a dozen operations can be replaced by a single versatile function. An example is the image expression evaluator 'eval', to which we return later. On the other hand, it should also be noted that years of study may lead to just a few statements. For example, studies to find the theoretical best approximation of the length of straight lines [25], or the curvature of shapes [26], result in a single statement or single operation, but one well founded.

The following gives an incomplete overview of the image processing functionality supported in ScilImage.

- Filters, including uniform, kuwahara, sigma, percentile, gauss and fuzzy derivatives, convolution with user defined kernel, various edge-detection filters etc. etc. Where possible the implementations provide filter size independent performance.

- Fast binary image and mathematical morphology operations, grey value mathematical morphology and threshold morphology, see below.
- Full Image arithmetics, logical functions, and versatile image expression evaluator, see below
- Local image processing operations on interactively pointed image patches ("lens").
- A number of segmentation and threshold tools, dynamic segmentation and path search operations with cost values, Freeman chain code handling.
- Geometric manipulations, such as rotate, panning, and alike.
- Mouse based image interactions.
- Display: 2D and 3D with programmable LUT, 3D slicing and volume rendering, automatic enhancements, settings, scaling, sizing, positioning, and dithering.
- Test pattern generation.
- Disk i/o (TIFF, PICT, TCL-Image, AIM, ICS). Image postscript generation.
- Image statistics.
- Graphics overlaid on image.
- Interactive and automated particle shape and density measurement, selection and analysis.
- Various threshold and segmentation methods.
- Fourier analysis, frequency filtering.

Blazing Bit-mapped Binary (BBB) Images

We present a few topics in more detail.

Many of the image processing functions are built on the experience with the previous TCL-Image [13] and ScilAim [14] packages. The representation of binary images in these packages, however, was in conflict with the class separation and class safety policy we advocate. In the new binary image processing library, binary images are represented as binary bitmaps, rather than as bit-planes in grey value images [27].

The new representation not only dramatically improves the execution speed, but is also much more efficient regarding memory requirements. It forms an appropriate format for document image processing.

In the bit-mapped binary data-structure 32 consecutive binary image pixels fit in a single computer word, allowing basic operations to operate on 32 pixels in parallel. Boolean operations, for example, are speeded up by a factor of 50. On common general purpose architectures they are executed well within real time [27].

A methodological advancement through applying logarithmic decomposition of structuring elements in mathematical morphology, results in erosions and dilations of convex structuring elements, which are an order of magnitude faster than the previously best known software implementations [27]. Independent of image content 17 iterations of an erosion on a 512*512 image, for example, complete within 60 ms on a Sun SparcII rated at 28 MIPS.

Threshold morphology or fuzzy mathematical morphology was introduced in [28] to make the morphology theory more robust against noise in the image. Rather than requesting that precisely 1 pixel in the structuring element is an object pixel for removal by the erosion, threshold morphology replaces it by a threshold value, t. For t=1 or t=N, threshold morphology is identical to a dilation or an erosion respectively. For all values in between it allows some tolerance in the data (noise) or the model (the shape of the structuring element). Threshold morphology improves the practical application of mathematical morphology. As all other mathematical morphology implementations in the package, threshold morphology uses the BBB-data structures and supports arbitrary shaped structuring elements.

Also the grey value morphological operations have been implemented in ScilImage. It has been shown in [29] that erosions and dilations with parabolic shaped structuring functions are the morphological equivalents of the linear Gaussian scale space filter. A practical advantage of parabolic structuring functions is that they are the unique rotational symmetric structuring functions that can be decomposed by x- and y- dimensioning, yielding an acceptable computing performance.

Image Expression Evaluator

In ScilImage, an image expression evaluator is a handy instrument for image generation, quick modification, and for developing an ad-hoc feel for the image contents. Providing separate expressions for 'add_image', 'multiply_image', and so on, would be cumbersome. In ScilImage this is solved through an image expression evaluator named 'eval', which accepts all C expression operators (and more). We illustrate its use by some examples:

```
eval "B = 3 * A + C / 3.5 * (D / 2 + 128)"
```

Thresholding an image by a certain value can be expressed by introducing some pixel dependent control as:

```
eval "B = (A<128) ? 0 : 255"
```

Where 'A', 'B', 'C', and 'D' are image names.

Not only pixel wise expressions are supported in the eval command, but also neighborhood operations using indexes are defined. The indexes can also be expressions.
The following command calculates the Laplacian of image 'A' and stores the result in 'B':

```
eval "B = 4 * A - A[1,0] - A[0,1] - A[-1,0] - A[0,-1]"
```

Where 'A' can either be a grey valued integer or grey valued floating point image.

Although the calculations done with 'eval' are a factor 5 to 15 slower than using compiled routines the advantages are obvious: i) intermediate results are calculated in high precision, only the final result is converted to the pixel format of the receiving image, and ii) the simple syntax makes it easy to use. Moreover, in the final application the expression can be programmed in C and added to the system as a compiled routine to benefit from improved execution speed.

5 Related Work and Use

In [30], Dobie and Lewis discuss a library of functions and data structures that provides an extensible framework for implementing image processing algorithms.

They emphasize the importance not to bother the user with different names, and use generic functions to handle an operation for all images classes, as we do. Their generic functions, however, do not employ an abstraction mechanism (such as function overloading). As a consequence the generic functions tend to become extremely complex as each function has to deal with all possible image types. Furthermore, a new image type requires a change of all generic functions.

Where they take automatic adjustment of output image types into consideration, they reject this principle claiming that this would confuse the user. Our experience in practice with expert and layman users indicates the contrary. Rather it is exactly this behavior that contributes to a logical view on image processing which facilitates the use of the system.

Elements of the current approach to the presented environment for image processing can also be found in [31, 8, 12], and to a lesser extend in [32, 33].

Visilog [31] resembles SciIImage the most. The extensible system also provides multiple interface layers including a programmable menu and dialog system and a C interpreter. Compared with the presented system, Visilog offers the programmer more flexibility in the design of user interface components at the price of introducing different interaction styles for the various software modules. As a consequence, adding new functionality is more complicated. At the bottom line, Visilog has a serious drawback compared to SciIImage in its price tags.

Another comparable extensible image processing system is Khoros [12], which offers the user a visual programming layer (cantata) similar to the corresponding layer of SciIImage. It is amazing that Khoros exploits little of the visual character of the VP layer, e.g. not showing icons which visually represent the corresponding task. A command line interface is present in Khoros, however severely restricted as it is not a standard programming language. A number of tools formalize how a user can extend the system, including consistent user interface components. A drawback of the task oriented nature of the system is that it is huge in size (ca. 100 Mb). Since each operation involves starting a new task to be fetched from disc, there is a heavy overhead in performing a sequence of operations. On the positive side, other programs are easily interfaced to Khoros, and execution of tasks can be distributed over a network. Khoros is in the public domain, widespread in use, and as such provides an image processing environment to many research institutions.

SciIImage is in operation now for 3 years. It has been in use on an international scale on many platforms, including all major Unix systems and the Macintosh (with MS-Windows underway). It is in use in universities, and research institutions, as well as in industry. On most sites SciIImage is in use for image processing research, development and teaching. The outcome of the image processing research is integrated into new versions of the SciIImage product.

Depending on machine architecture, the complete package requires between 7 and 10 Mb of disk space. The SciIImage executable is ca. 900Kb on Macintosh and MsWindows. On UNIX X workstations the executable is about 1.2 Mb or 2 Mb on CISC and RICS architectures respectively.

Projects

SciIImage has proven especially effective as a vehicle for research collaborations. Below we present a selection of some of the large-scale projects using SciIImage:

SPIN-3DIA The Dutch national SPIN-3DIA research program aims at stimulating image processing research of three dimensional microscopical images [34]. The research program involves 7 institutes. In the program methods have been designed for the restoration of microscopical images, for the automated registration of subsequently captured images, as well as various methods for 3D-segmentation, for the proper measurement of properties such as length, curvature and methods for the visualization of 3D-data sets. Apart from methodology, also practical applications in the domain of biological microscopy have been solved, all based on ScilImage.

ROCKI ScilImage has been successfully utilized to serve as a common development platform in the ESPRIT II project 5376 ROCKI: Raster to Object Conversion aided by Knowledge Based Image processing.

The main objective was to specify and experiment with a common toolkit for the high level interpretation of office documents and cadastral based maps. To reach this objective, techniques have been developed to convert the raster representation of scanned documents into a description of objects, their attributes and relationships.

CHIC ScilImage is effectively being used in ESPRIT III 6765 CHIC: Component Heuristics and Improvements for Copiers.

The prime objective of CHIC is to realize a demonstrator to show feasibility of a digital copying device, based upon component oriented image processing techniques. The availability of a digital copying device gives an advantage over the conventional analog copiers as the copying process can separately optimize for text, images and graphics recognized in a document.

Pathology Image Processing Environment At the institute of pathology of the Free University of Amsterdam, ScilImage is tailored to function as an image processing environment for microscopical analysis in pathology. Applications include, automated DNA-measurement [35], tissue epithelium estimation [36], and recognition of mitosis [37], for tumor diagnostics.

Model driven image processing The University of Amsterdam is involved in a national research project concerned with "model driven image processing". Most of the participants (five universities in The Netherlands) are using ScilImage as a development tool facilitating the exchange of research results. In this project, our group concentrates on building a framework for image understanding and interpretation in medicine, geography and robotics.

Videophone ScilImage is being used, by the Faculty of Electrical Engineering of the Technical University Delft, in a large-scale interdisciplinary project concerning the development of videophone technology.

CommonViewing Philips Medical Systems have integrated the ScilImage environment into their 'Common Viewing' toolbox, a common product interface and viewing environment for medical imaging, which serves as a unified reusable software base for several current and future workstation products of PMS.

As an integrated part of Common Viewing, ScilImage can be used in two ways: as a development environment for new versions of PMS workstation products (allowing experimentation with image processing algorithms and with user interface approaches) and, optionally, as a 'hidden' image processing engine in products.

The ScilImage interface layers have been re-styled by PMS to conform with the Common Viewing look&feel. Although ScilImage and Common Viewing have different image datastructures, a special interface allows transparent functionality of both systems on all images.

CAGDIS In the request for the European Community Mobility project 'Computer Analysis for Geometrical Description of Imaged Scenes', ScilImage is proposed as the common platform for interchange and comparison of results by the participating groups from Grenoble (France), Arco Felice (Italy), Linkoping (Sweden), Vienna (Austria) and Amsterdam (The Netherlands).

ScilImage was chosen for the flexibility with which one can customize the package with the latest results, while maintaining the commonality of the standard routines.

For improved support ScilImage (except for the VP layer) has been commercialized by the Center of Pattern Recognition (CBP), a Dutch institute in which the Technical University Delft (TUD), the University of Amsterdam (UvA), and the Dutch Institute for Applied Physics (TNO) participate.

6 Discussion and Conclusion

In this paper it is argued that, if an image processing environment is to meet the requirements of different type of users, and multiple research institutions, the time is felt right, to specify a portable and very flexible setup to allow for ongoing changes in software methodology. As it meets these requirements, we advocate the use of a machine independent, multi-layered software environment, combined with an object oriented infrastructure for image processing.

To meet the requirements of different type of users, we have designed and implemented an interactive image processing environment consisting of simultaneous interface layers. The system offers several advantages, the main issues we list here.

The library handler provides easy integration of external libraries of various engineering fields, such as image processing, databasing, graphics, and statistical analysis. The C interpreter SCIL furnishes the application developer with the flexibility of a high level programming language. A command expander hides the C level and provides the experienced user a high level design tool. The menu and dialog generator provides a simple to use point and click interface to help the novice user locating functions. To combine user-friendliness with programmability also a visual programming layer allows non-programmers to construct simply structured but non-trivial programs by connecting icons on a worksheet.

Since the environment is responsive to lay users as well as experienced developers, the gap between development and use of a system is resolved.

All the interface layers are generated from specifications in the Command Description File. Therefore, the environment allows for easy tailoring the user interface to a particular application or end-user group. In addition, fully developed applications can be compiled and used as stand-alone tasks or added to the system.

In the system design emphasis is given to image class separation and the importance to represent images in a format most efficient for processing. This leads to the conclusion that an object oriented image infrastructure can and should generalize actions common to all image classes. The image infrastructure takes care of ROI processing, class safety, consistency, and standardized checks on function arguments, rather than making this the concern of the programmer.

Moreover, the user is given freedom of output image designation, with the infrastructure automatically taking care of the adjustment of the output image in all situations. The main advantage is that this leads to a dynamic and logical view on image processing rather than a static and algorithmic one, facilitating the use of the system.

In the image infrastructure, function overloading is used to communicate with different image classes in an abstract way. It supports creation of new classes and image sub-classing, where new images automatically inherit the consistent behavior common to all image classes. To our knowledge no present image processing system supports this level of extensibility.

The natural classification of images as well as the object oriented infrastructure, would benefit from a C++ implementation. The availability of a C++ interpreter would, therefore, be an improvement to the total environment but is difficult to realize.

The ease with which (newly developed) software can be exchanged between users has made SciIImage especially fruitful as a common development vehicle in research collaborations. With so many researchers developing new algorithms, the image processing functionality of SciIImage is growing at a more accelerated pace than a single research institution could achieve.

Finally, based on it widespread use, we conclude that SciIImage is not limited to a single research setting since it meets the demands of portability, maintainability, and flexibility necessary to function as a general vehicle for use and development of image processing software.

```
Prices and further information about SciIImage can be obtained
through:
email: scil@tpd.tno.nl

regular mail: TNO Institute of Applied Physics
PO Box 155
2600 AD Delft
The Netherlands
```

References

[1] B.G. Bailey and R.M. Hodgson. VIPS: A digital image processing algorithm development environment. *Image Vision Computing*, 6(3):176–184, 1988.

[2] A.G. Sleigh and P.K. Bailey. DIPOD: An image understanding development and implementation system. *Pattern Recognition Letters*, (6):101–106, 1987.

[3] J. Mundy et. al. The image understanding environment program. In *Proceedings of the CVPR*, pages 407–415, 1992.

[4] B. Zavidovique, V. Serfaty, and C. Fortunel. Mechanisms to capture and communicate image processing expertise. *IEEE Software*, pages 37–50, November 1991.

[5] W. Havens. PIPS: A portable image processing system. In *Proceedings of the 1986 Workshop on Visual Languages*, pages 125–133, 1986.

[6] K. Preston Jr. Image processing software. a survey. In L. Kanal and A. Rosenfeld, editors, *Progress in Pattern Recognition*, volume 1, pages 123–148, 1981.

[7] P. Gemmar and G. Hofele. An object oriented approach for an iconic kernel system IKS. In *Proceedings of the 10th ICPR*, volume 2, pages 85–90, 1990.

[8] G. Fisher. An overview of a graphical multilanguage applications environment. *IEEE Transactions on Software Engineering*, 14(6):774–786, June 1988.

[9] M. Kaneko and K. Sakaue. Design and implementation of SPIDER- a transportable image processing software package. *Computer Vision, Graphics, and Image Processing*, (23):273–294, 1983.

[10] *IRIS IPS/68k Reference Manual. Kontron Image Analysis Systems 1.1*, 1985.

[11] F.C.A. Groen, R.J. Ekkers, and R. de Vries. Image processing with personal computers. *Signal Processing*, (15):279–291, 1988.

[12] J.R. Rasure and C.S. Williams. An integrated data flow visual language and software development environment. *Journal of Visual Languages and Computing*, pages 217–246, September 1991.

[13] G. van Antwerpen and R.J. van Munster. *TIPS Users Manual.* Institute of Applied Physics-TNO, Delft, The Netherlands, 1987.

[14] T.K. ten Kate, R. van Balen, A.W.M. Smeulders, F.C.A. Groen, and G. den Boer. SCILAIM: A multi-level interactive image processing environment. *Pattern Recognition Letters*, (11):429–441, June 1990.

[15] J. Graham, C.J. Tailor, D.H. Cooper, and R.N. Dixon. A compact set of image processing primitives and their role in a succesful application program. *Patteren Recognition Letters*, (4):325–333, 1986.

[16] A.R. Jones and E.R. Steinberg. Interface design: An educational perspective. *Journal of Systems and Software*, 11(3):199–204, 1990.

[17] R.van Balen, T.K. ten Kate, D. Koelma, and A.W.M. Smeulders. SCIL: A machine independent multi-layered environment for use and development of software. In Prep.

[18] R. van Balen and A.W.M. Smeulders. A flexible datatypes infrastructure for image processing. In prep.

[19] D. Koelma, R. van Balen, and A.W.M. Smeulders. SCIL-VP: a multi-purpose visual programming environment. In *Proc.eedings of the 1992 ACM/SIGAPP Symposium on Applied Computing*, volume 2, pages 1188–1198. ACM Press, March 1992.

[20] N.C. Shu. *Visual Programming.* Van Nostrand Reinhold Company Inc., 1988.

[21] S.K. Chang. Visual Languages, a tutorial and survey. *IEEE Software*, pages 29–39, January 1987.

[22] B.A. Myers. Taxonomies of visual programming and program visualization. *Journal of Visual Languages and Computing*, pages 97–123, March 1990.

[23] J. Piper and D. Rutovitch. Data structures for image processing in a C-language and UNIX environment. *Pattern Recognition Letters*, 3(2):119–129, 1985.

[24] J. Piper and D. Rutovitch. An investigation of object-oriented programming as the basis for an image processing and analysis system. In *Proceedings 9th ICPR*, pages 1015–1019, November 1988.

[25] L. Dorst and A.W.M. Smeulders. Best linear unbiased estimators for properties of digitized straight lines. *IEEE transactions on Pattern Analysis and Machine Intelligence*, (8):276–282, 1986.

[26] M. Worring and A.W.M. Smeulders. Digital curvature estimation. in press for CGVIP:Image Understanding.

[27] R. van den Boomgaard and R. van Balen. Methods for fast morphological image transforms using bitmapped binary images. *CVGIP: Graphical Models and Image Processing*, 54(3):252–258, May 1992.

[28] R. van den Boomgaard. Threshold logic and mathematical morphology. In *Progress in Image Analysis and Processing*, 1989.

[29] R. van den Boomgaard. The morphological equivalent of the gauss convolution. *Nieuw Archief voor Wiskunde (in English)*, November 1992.

[30] M.R. Dobie and P.H. Lewis. Data structures for image processing in c. *Pattern Recognition Letters*, (12):457–466, August 1991.

[31] *Visilog User Manual.*

[32] C. Upson et. al. The application visiualization system: A computational environment for scientific visualization. *IEEE Computer Graphics & Applications*, pages 30–42, July 1989.

[33] *KBVISION User Manual.*

[34] I.T. Young and J. van Renswoude. SPIN Three-Dimensional Image Analysis. Proposal to SPIN, January 1988.

[35] A.W.M. Smeulders and T.K. ten Kate. *Accuracy of Optical Density Measurements of Cells. 1: Low resolution.* PhD thesis, Free University Amsterdam, 1990.

[36] N.W. Schipper, A.W.M. Smeulders, and J.P.A. Baak. Evaluation of automated estimation of epithelial volume by image processing applied to ovarian tumors. *Cytometry*, (8):345–352, 1987.

[37] T.K. ten Kate, J.A.M. Belien, A.W.M. Smeulders, and J.P.A. Baak. Method for counting mitosis by image processing in feulgen stained breast cancer sections. *Cytometry*, (14):241–250, 1993.

Vision as Process: Integration and Control of a Real Time Active Vision System

James L. Crowley
LIFIA - IMAG
46 Ave Félix Viallet
38031 Grenoble, France

and

Henrik Christensen
Laboratory of Image Analysis, Aalborg University
Fr. Bajers Vej 7 D
DK-9220 Aalborg, Denmark

ABSTRACT

This paper presents a software skeleton system for experiments in integration and control of real time active vision systems. This skeleton has been constructed by a consortium of six laboratories as part of a long term basic research investigation of issues of integration and control of real time vision.

The first chapter describes the problems of integration and control and the context for their investigation. The second chapter reviews existing architectures for the integration of vision systems. Concepts from each of these architectures are adapted into a proposition of an architecture composed of a distributed set of modules.

The third chapter presents the SAVA III skeleton system which has been developed to support our experiments in integration and control. An overview of the skeleton system is presented, and then the communication mechanisms are described. The use of a rule based interpreter based on CLIPS 5.1 permits each module to send messages which can define rules and functions within the other modules, essentially allowing modules to program each other. The rule interpreter is used to construct the control component for the processing cycle for each module. The interpreter is also used for the control part of pre-attentive demons which detect perceptual events within the modules.

The SAVA system is illustrated with a description of a recent demonstration of a visual navigation system constructed in the SAVA skeleton. The components of the visual navigation system are presented and details are given concerning the use of stereo for obstacle detection, and the use of landmarks for updating the estimated position.

1 Introduction

Active vision is the use of controlled camera motion and control of processing to simplify and accelerate visual perception. This sub-discipline has grown steadily from ideas expressed in the mid 80's by Bajcsy [Bajcsy 88], Aloimonos [Aloimonos et al. 88], Ballard [Ballard 91] and others, to what now appears to be a "paradigm shift". Part of this shift has been the recognition that a vision system can not be designed in isolation of the task. Vision must serve a purpose, and in particular should be designed in the context of the perceptual needs of the task and the environment. This leads to a view of a vision system which operates continuously and which must furnish results within a fixed delay. Rather than obtain a maximum of information from any one image, the camera is an active sensor giving signals which provide only limited information about the scene. The development of robotic camera heads has lead to the possibility of exploiting controlled sensor motion and control of processing to construct continuously operating real time vision systems.

At the same time, research in applying artificial intelligence techniques to machine vision led to an emphasis on the use of declarative knowledge to control the perceptual process. Systems such as the Schema System [Draper et al. 89] developed a black-board architecture in which multiple independent knowledge sources attempted to segment and interpret an image. A major problem in such systems is control of perception. Such systems emphasise explicit representation of goals and goal directed processing which direct the focus of attention to accomplish system tasks. It has not been obvious how such a knowledge based approach to control of attention could be married to a real time continuously operating system.

In July 1989, the European Commission funded a consortium of six laboratories to investigate control of perception in a continuously operating vision system[1]. The consortium partners set out to build a test-bed vision system for experiments in control and integration. An experimental test-bed system was constructed which integrates a 12 axis robotic stereo camera head mounted on a mobile robot, dedicated computer boards for real-time image acquisition and processing, and a distributed system for image description. The distributed system includes independent modules for 2-D tracking and description, 3-D reconstruction, object recognition, and control. On March 18 1992, a fully integrated continuously operating vision system was demonstrated to the European Commission using this test-bed. This paper reports on the development of this system and the research which the system makes possible in control of a real-time vision system. A more complete description of the results of the project may be found in the book [Crowley –Christensen 93].

---

[1]The Partners in project ESPRIT BR 3038 are Aalborg University (DK), University of Surrey (UK), KTH, The Royal Institute of Technology (S), University of Linkoping (S), LTIRF-INPG (F) and LIFIA-INPG (F). The project has been continued under BR 7108 with the addition of University of Genoa (I).

1.1 The Project Vision as Process

The starting point for the project "Vision as Process" was the demonstration of an integrated vision system capable of continuous real-time operation. It was quickly realised that such an ambition raises two problems:

1) The technical problem of integrating processes which model the environment in terms of descriptions which are qualitatively different.

2) The problem of controlling the "attention" and processing of a continuously operating system.

Concerning the first problem, different robotic tasks require different kinds of descriptions of a scene. Such descriptions can include 2D image description, 3D scene descriptions and symbolic labelling of the components of a scene. Such processes are complementary and mutually supportive. A framework is required which would permit the integration of multiple vision processes. This can be considered an "engineering" problem.

The second problem is both subtle and fundamental. Most of the algorithms used in vision have computational costs which depend on the quantity of data. In the best cases the relation is linear, but in many cases it is quadratic, cubic, or even exponential. Real time response requires that the processing time for any part of the system is limited. This requires that the amount of data considered during each processing cycle be bounded which raises the problem of which subset of the available data the system should attend during each cycle. This is part of the larger problem of controlling perception. General purpose real time vision system requires a solution to this problem.

These observations let the consortium to develop a long term work plan with both an engineering and a scientific goal. These are:

Engineering Goal: Develop techniques for integrating cyclic real time processes for description of a scene in terms of 2D images, 3D structure and labelled objects using active control of a camera head.

Scientific Goal: Develop methods (and eventually a theory) of control of attention in perceptual processes.

With these twin goals, the consortium has developed a long term plan leading to the demonstration of methods for the construction of integrated continuously operating vision systems, and the elaboration of a theory for the control of such systems.

1.2 System Integration and Control

When the VAP project was conceived in 1988, only a small number of vision systems were capable of performing symbolic interpretation, and they were designed for interpretation of single (static) images. The well known examples included VISIONS [Hanson-Riseman 78], ACRONYM [Brooks 81], and 3DPO [Bolles-Horaud 84].

Most work on analysis of image sequences had been carried out on pre-recorded images and the level of description was almost entirely parametric. i.e., systems could describe regions or features with independent motion in terms of their image or 3D-velocity. A review of the state of the art is provided by Huang [Huang 83]. Continuous and real-time observation of a dynamically changing scene involves more than motion interpretation. A continuously operating vision process must be able to limit processing to a small subset of the data available from visual sensors, and to adapt its processing mode dynamically in response to events in the scene and requirements of the task.

1.3 The VAP Hypotheses

From its earliest meetings, the VAP consortium agreed that vision should be studied in the context of its purpose, i.e. its use by other processes. Without dedicating to any specific application, this implies that visual processing can be controlled to concentrate on the subset of visual information which is considered relevant to the current goal as defined by a user process. In addition, the consortium recognised the ability to exploit coherence in the dynamic evolution in a scene. In a continuously operating system, temporal context permits changes in the scene to be predicted and computational resources to be directed to confirm expectations. This implies that tracking is basic operation within a continuously operating system.

The goal of the VAP project is to demonstrate that a vision system must be designed as a continuously operating "process". To demonstrate this principle, the consortium has designed a six-year research program to develop techniques to interpret a dynamically changing, quasi-structured environment. These techniques exploit goal directed focus of attention involving controlled sensor motion and control of processing. Processing is directed by goals which change dynamically in reaction to the needs of the perceptual tasks and to events in the scene.

The following section reviews some previous approaches to integration and control. This previous work establishes the set of concepts and "prior art" from which the design of the VAP skeleton system draws.

2 System Architectures for Integration and Control

A suitable system architecture is required for experiments in integration and control of a continuously operating system. In this chapter, we review previous approaches to architectures of vision systems. From this review, we argue that flexible integration may be achieved through use of a standard module architecture, replicated at each of the levels in the system. Such a standard module architecture is described in more detail in chapter 3.

2.1 The Reconstruction Approach

A popular approach for structuring a vision system has been proposed by Marr [Marr 82]. Marr argues for a system defined around a hierarchy of representations: images, The primal sketch, the 2.5-D sketch (viewer centred depth map), 3-D map and symbolic description. In this model, processing is organised as sequential processes in which information flows up through the levels. Processing is data-driven, in the sense that recognition and description are based on descriptions constructed at the lower levels in the system. This model is computationally demanding and it has proven difficult (if not impossible) to provide image descriptors that are sufficiently robust to allow characterisation of all phenomena in a natural environment.

The Marr processing model may be termed a *reconstruction approach* as it aims at a full reconstruction of the environment. The processing model is purely data driven, and thus poses a problem in terms of computational resources. The Marr model assumes all processing may be carried out as a sequential process. This implies that a module uses the representation(s) at the level just below as a basis for its processing and the result is stored in the next higher representational level. The interfaces are consequently well defined. A simplified model of the processing is shown in figure 1.

In terms of representations, this processing model implies that information needed to perform recognition and interpretation of settings must be available as part of the 3D model. i.e., a diverse set of descriptors must be tagged onto the 3-D model representation to facilitate recognition and description. This duplication of information up through the system and the unavailability of pixel level primitives can pose a problem in terms of model size and maintenance over time.

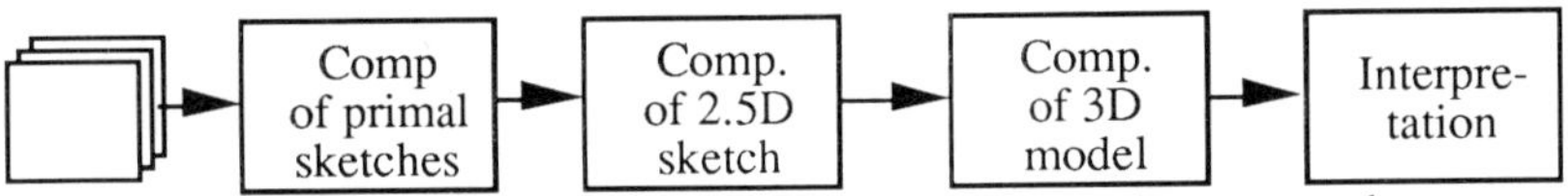

Figure 1 A processing model for the reconstruction approach.

2.2 The Non-Committal Approach

In the VISIONS system [Hanson-Riseman 1978] data are stored on a blackboard, or common storage area and processing is performed in parallel by a number of "knowledge sources". All modules in the system can access information at any of the representational levels. This implies that information does not have to be replicated up through the system to be made available for recognition procedures. A simplified model of the non-committal approach is illustrated in figure 2.

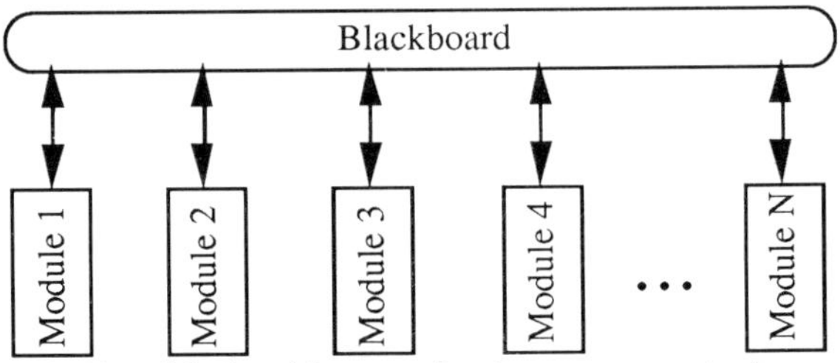

Figure 2. Simplified architecture for the non-committal approach.

In this architecture, each of the modules in the system has control information that specify the information that must be available before the module may carry out its task. In addition, it has control information that specifies the information which may be provided by the module. Through use of a control executive, it is possible to perform both goal directed and data driven processing through use of this control information.

This processing model imposes few constraints on the representations used in the systems and it is simple to add new modules to the system. The common blackboard is a potential problem for a continuously operating system. All information generated and used by the system passes through the blackboard. It might thus pose a problem with respect to information band-width. To indicate the amount of information, which may be generated directly from images one may look at the Image Understanding Architecture, described by Weems et al. [Weems 1989]. In that system, the storage reserved for intermediate representations is 4 GB and the system is only aimed at analysis of single images. Such an architecture will require extensive use of special purpose hardware, in particular when applied in the temporal context.

2.3 The Purposive Approach

Introduction of goal directed operation and use in a limited and well defined domain of application allow synthesis of a vision system which is composed of a set of specifically engineered modules. Such modules may be designed to be computationally well behaved, in the sense that the computational complexity is bounded and often robust representations can be provided. This approach to construction of vision system has been promoted by Bajcsy [Bajcsy 88], Ballard [Ballard 91] and [Aloimonos et al. 88]. Although the approach exploits specific vision modules, Bajcsy has tried to enumerate a set of modules that might form a general purpose system. The use of dedicated modules is a way to provide robust information and computationally tractable techniques. Well known examples of dedicated modules used for robot navigation are optical flow modules that can compute the position of the focus of expansion for the optical flow field, and modules which can compute the time to contact from motion in an image sequence.

This approach to system construction is termed *the purposive* or *the animate* approach. It is envisaged that the construction and analysis of specific modules will gradually provide insight that will allow definition of modules applicable in general vision systems. The convergence towards a standard set of modules through analysis of diverse application domains might provide valuable insight, but it is not obvious that convergence will be achievable.

In the purposive approach, the exploitation of information is task driven and may be very different from one task to the next. The basic system architecture should thus be flexible and facilitate dynamic change of the information flow. In practical systems a number of modules may exploit the same representation and once a system has been defined an analysis of the representational requirements may point to the definition of a set of standard representations. Given present state of the art no such general representations are known.

A purely purposive approach to vision rejects most of the established techniques.

Modules are to be built from scratch whenever a new type of information/representation is required. There is consequently a concern that from this approach little insight will be gained in terms the general vision problem.

2.4 The Vision as Process Architecture

During the first year, the consortium "Vision as Process" addressed the problem of design of an architecture which would meet the following criteria:

1) Continuously operating.
2) Integrating description of the environment with 2D measurements, 3D models and recognition of objects.
3) Capable of supporting diverse experiments in gaze control, visual servoing, navigation and object surveillance.
4) Dynamically reconfigurable as the task changes.

The result was the design of a distributed test-bed system composed of independent modules. Modules may communicate by message passing over a central message server, or by dedicated "high-band width" channels. Systems can be composed from sub-sets of the available modules for individual experiments.

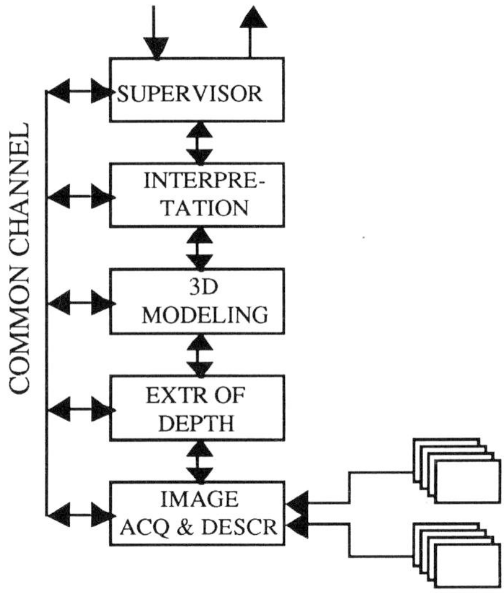

Figure 3. The VAP system architecture

The architecture adopted by the consortium is shown in figure 3. The system has a data flow part, which is similar to the Marr processing model. It should be noted that the data flow is not only bottom-up but may also be top-down. Top-down expectations (derived from the present set of goals and contextual information) can be used to direct/control processing at lower levels, while detected event at the same time can drive a reconstructive mode of processing. The VAP architecture contains a common communication channel that allow communication between any pair of modules in the system. This communication channel may be used both for investigation of a non-committal processing model, and for investigation of purposive systems, as the component processes in the system in figure 3 may either general purpose or dedicated.

This architecture imposes few constraints on the design of component modules and it provides flexibility for the investigation of system level control issues. Initially it was envisaged that the main flow of data would exploit the communication links between adjacent modules, while only control information would be communicated through the common channel. During the execution of the project it was realised that a more flexible processing model was needed to make computations both efficient and robust.

Having selected a distributed architecture composed of modules, the consortium turned its attention to the design of a common component for each module. At the very least, this standard module must provide the communications interface. It was soon observed that scheduling was a basic to continuous operation and a cyclic scheduler was provided which calls the procedures which implement each phase of computation. The phases of operation included phases for integration of new data and phases for control of processing.

In order to obtain temporal context, the consortium drew on previous results in image tracking. A tracking architecture was defined composed of the phases predict-match-update [Crowley et al. 88], [Granum-Christensen 88] based on techniques used in the control community since the early sixties [Kalman 60]. The architecture is shown in figure 4.

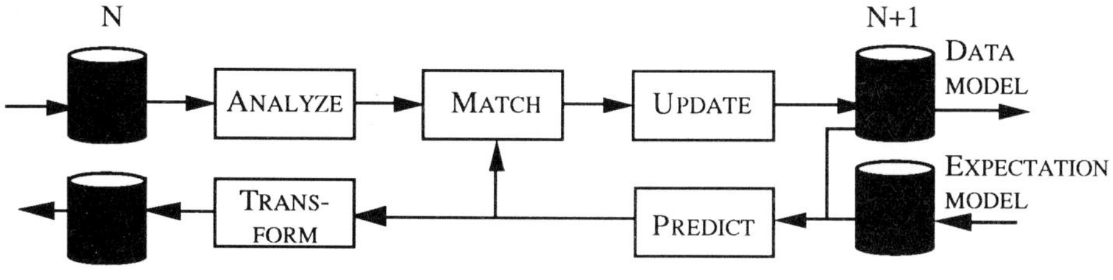

Figure 4. Basic Predict-Match-Update cycle for the module architecture.

The *analysis* block, in figure 4, is responsible for the frame by frame analysis, which generated a set of geometric primitives (or tokens). The correspondence with information in the temporal context is performed in the *match* block. To simplify matching the information in the temporal context is used in a prediction of the expected content of the next frame. Once correspondence has been established the information contained in the internal models must be updated to reflect the new information contained in the new frame. Once updating has been carried out the cycle may start all over again.

As a model at level N+1 is used for prediction of primitives in the next frame, the predictor may also be given other types of input which can be used for guidance of processing. Introduction of goal derived information into the model at level N+1 will consequently allow top-down/attention based control of processing. A prediction may be transformed into a representation that is compatible with the one used the level below, so that it may drive processing at the level below. This flexibility facilitates investigation of different control strategies.

2.5 Control Issues

Construction of an operational system includes issues in control to ensure satisfaction of user defined goals. Goals are widely recognised as a fundamental component of intelligent systems. The consortium initially defined a set of general goal commands :

search(X):	Is X present in the scene?
find(X):	Where is X, given X has been identified earlier?
relate(X,Y):	What is the spatial relation between X and Y?
describe(X,Y):	Determine property Y for object X
watch(X):	Allocate resources for notification of changes for X
track(X):	Maintain a dynamic description of object X.

This set of goals defines the user level interface to the system. Based on the potential goals, the system must be able to allocate its resources for optimal satisfaction of the concurrent goal(s).

A number of approach to goal directed processing have been reported in the vision literature. Most of these efforts include use of a cyclic process, in which data received are matched against expectations. Depending on success or failure in the matching process an updating or event detection process is used to drive the next cycle of processing. An example of such a cyclic process is described by [Tsotsos 87] for the ALVEN system.

When the VAP effort was initiated the use of production systems and reasoning under uncertainty appeared the most promising in terms of providing insight into the problem of the cycle of planning, sensing and interpretation. These tools have been incorporated into the system. In system level control, externally defined goal commands are translated into actions by rule based planning. Planning generates the sequence of state transitions (actions and their parameters) expected to allow completion of a goal. These actions are then executed by one or more system modules. The internal handling of such actions is an issue that is resolved for each of the modules. A skeleton system constructed by the consortium provides the framework for experiments in control and coordination of visual processes.

3. The SAVA III Skeleton System

In order to perform experiments in control and integration of a continuously operating vision system, the VAP consortium constructed an empty "skeleton" system. This skeleton was then provided to partners so that they could "fill in" the functional parts needed for their experiments[2]. This system was named SAVA, for the french acronym "Squelette d'Application pour la Vision Active". The SAVA Skeleton system provides a standard module with communication and interface components that permit an experimenter to construct and run distributed real time vision system.

The structure of SAVA has evolved with our understanding of the problems of integration and control. The original SAVA system was released at month 12 of VAP-I. Experiences during the second year of VAP-I brought out a number of shortcomings in the design. A team composed of people from AUC, KTH and LIFIA designed a revised version, SAVA II, which was released at the month 24 milestone. An intense integration effort was performed in preparation of the month 33 integrated demonstration, with software and hardware contributions from all VAP partners integrated into SAVA II. Modifications in communications and interface design, as well as a large number of small improvements, led to release of SAVA 2.4 in March 1992.

Experience with SAVA II has shown the importance of demons for combining purposive and event driven control of perception. This led to the desire for an interpreter for demon functions. In addition, programming control experiments in SAVA II was a somewhat difficult task. Control knowledge was embedded in procedural code and thus hard to understand or change. It was decided to design a control system based on an interpreter for declarative expressions of the control logic. From these two needs emerged the idea of using the CLIPS 5.1 rule interpreter for the control component and the demon interpreter within each module. CLIPS 5.1 is written in C and is provided with the full source code. As a result, it was extremely easy to integrate the SAVA modules into the CLIPS environment.

A new version of the skeleton system, SAVA III, has been created based on the principle of interpreting control information. In SAVA III, most of the procedures for processing and communication are written as C procedures and explicitly declared to the CLIPS 5.1 rule interpreter. Rules and functions are then written using these procedures. The basic processing cycle is built as a sequence of states with transitions managed by rules. The processing performed within a state can be easily changed based on either perceptual events or external commands. Because the control rules are interpreted, the control sequence for a module may be changed dynamically, without re-compiling a module. It is even possible for a module to send another module function definitions as ASCII messages, using the CLIPS deffunction facility. The rule based scheduler is particularly useful for the implementation of demons. Demons may be programmed as rules which react to the contents of the model as well as to external messages.

In addition to the changes in the control part, a major effort has been made to add the possibility of synchronised operation to the modules. SAVA is a soft real-time system, distributed over a set of workstations operating under UNIX. At some time in the future, we intend to port SAVA onto dedicated hardware running under a real time programming environment. However such systems are relatively difficult to program and debug. The use of UNIX and distributed processing permits the VAP-II project to perform

---

[2]The incompatibility of successive releases of MOTIF have created problems for portability and have cost the consortium considerable time and money.

experiments with a reasonable effort.

A synchronisation system has been built into SAVA III in order to compensate for uncertainties in communication and execution time for distributed modules. A synchronisation module provides other modules with a universal time reference. In this way, all information that is processed or communicated is time-stamped, permitting an estimate of dynamic processes to be observed or controlled.

The following sections present a detailed description of the components of the SAVA III system. It first gives a brief overview of the components of the skeleton system and its standard module. It then describes processes for interpreting messages using a rule based interpreter, and the design of "demon" processes that perform pre-attentive detection of events. A description of the rule based control of a module is presented, followed by a description of the synchronisation of modules.

3.1 Overview of the SAVA III Software Skeleton

The SAVA skeleton system is composed of the following components:

1) A launcher program that permits the user to assign modules to processors and to initiate operation.

2) A distributed mailbox system that is launched on the different processors to establish a communications system and to launch the component processes.

3) A library of communication procedures for modules. This library include procedures for communication by message as well as procedures for dedicated high band-width communication between processes.

4) A skeleton module structure built around a scheduler.

5) A set of graphical man-machine interfaces.

The SAVA system provides mailbox communication for data, control and acknowledgements, as well as a procedures for dedicated high-band-width channels between modules. Messages include formatting information that permits the message passing system to pack and unpack messages.

Visual perception is performed within processes imbedded in copies of the SAVA "standard module". The SAVA III standard module is shown in Figure 5. The standard module is composed of a number of procedures (shown as rectangles) that are called in sequence by a scheduling process.

A SAVA module repeatedly executes a cycle in which it:

1) Acquires new data.
2) Transforms this data into an internal representation.
3) Makes predictions from its internal model.
4) Matches the predictions with the transformed data.
5) Uses the match results to update the internal model.

6) Executes demons to detect perceptual events within the internal model.

This cyclic process is executed by a rule-interpreter. Each phase corresponds to a state in which a particular operation is performed. At each state transition new messages which have arrived on the mail box channel are read and processed. Such messages may change the procedures that are used in the process, change the parameters that are used by the procedures, or interrogate the current contents of the description that is being maintained.

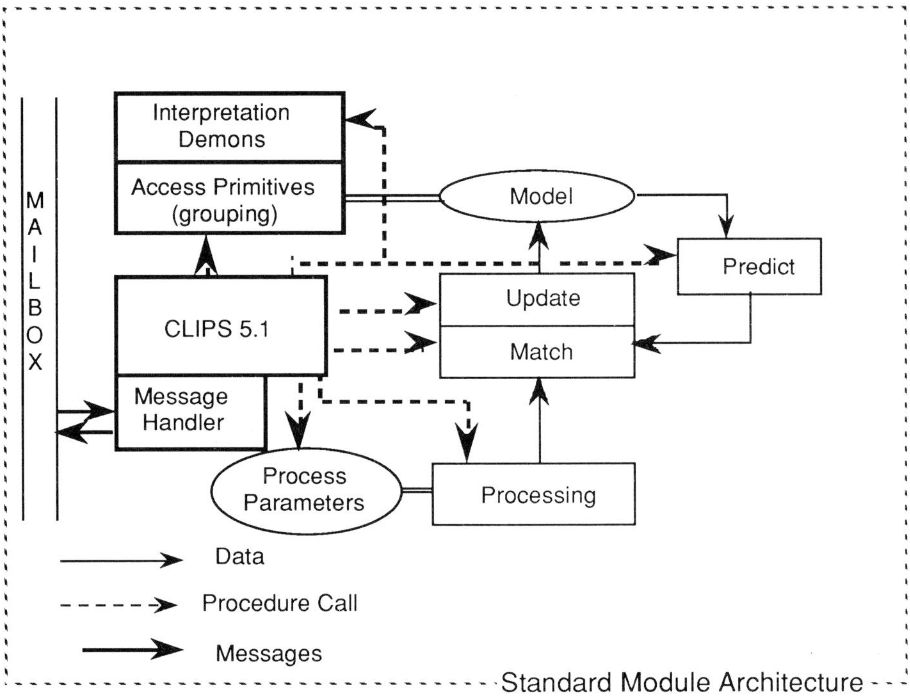

Figure 5 Architecture of a Standard Module in SAVA III

The cyclic process within a module is managed by a control token placed on the working memory. This control token is a simple list in which the first atom is the word "phase" and the second is the name of one of the phases: {get-data, transform, predict, match, update, messages, demons}. The definition of theses phases is as follows:

get-data	Acquire a new observation
transform	Transform the data to the internal representation

predict	Predict the contents of the observation
match	Match the prediction to the observation
update	Update the model using the correspondence of the prediction and the observation
demons	Execute a set of automatic procedures for event detection.

At the end of each cycle, the scheduler executes a set of demons. Demons are responsible for event detection, and play a critical role in the control of reasoning. Some of the demon procedures, such as motion detection, operate by default, and may be explicitly disabled. Most of the demons, however, are specifically designed for detecting certain types of structures. These demons are armed or disarmed by recognition procedures in the interpretation module according to the current interpretation context.

In the SAVA III system, the procedures of a module are made explicitly available to the CLIPS 5.1 rule based interpreter. This includes the original SAVA II scheduler, so that the system is upwards compatible. In addition to acting as a scheduler, the rule interpreter is also used to define the control part of demons and to interpret messages from other modules.

3.2 Communications Between Modules.

Modules communicate control, data requests, reply and synchronisation information using message passing based on Unix Sockets. The SavaSend() function is used to send mailbox messages to other modules. A SavaSend command contains three fields:

Header:	The destination and type of message.
Format:	An ASCII description of the message format.
Body:	The message including commands and parameters.

The destination is a symbolic name for another module. The types of message may be control, acknowledge or data. All message exchanges are initiated by a control message. The format string is transmitted with the message and is used both to encode and decode the message. In this way a change in message protocol may be made with a minimum of difficulty. This format string can contain conversion directives like %d, %f, and %c, based on the C language printf protocol. We have added conversion directives for sending arrays, structures and images.

Many SAVA functions accept a variable number of arguments. Furthermore, the type of these arguments is unspecified. These functions accept a fixed number of normal arguments, followed by an arbitrary number of arguments of unknown type. The last normal argument is a format string which describes the arguments following it.

Large data structures may be communicated between modules using dedicated sockets. Communication of dedicated channels are performed by the functions SavaRead and SavaWrite. As with mail-box, messages, high band-width channel messages are encoded with an ASCII format directive which is transmitted with the message. High band-width channels in SAVA are faster than message passing because the channels provide a direct connection. No intermediate routing is necessary.

Messages passed through the mail box communication system are interpreted by the rule interpreter. New messages are transformed into working memory elements by the function “checkmessage”. Checkmessage creates a list in working memory composed of the keyword “message” followed by the name of the sender, a message keyword and and ASCII string with the body of the message. Checkmessage assures upwards compatibility with message types that were defined in SAVA II which have not be transformed to SAVA III.

The checkmessage function is executed at the end of each phase of the standard module. For example, the transition from match to update is performed by the rule “update-phase” :

```
(defrule update-phase
    (declare (salience -100))
    ?p <- (phase match ?c)
=>
    (check-message)
    (retract ?p)
    (assert (phase update ?c))
)
```

The result of check-message is a list of the form:

```
(message  ?sender ?command ?body)
```

If ?command string is tested to determine how the message should be interpreted. If command corresponds to one of the CLIPS keywords “deffunction” or “defrule”, then ?body is interpreted by the CLIPS function BUILD. This is permits external modules to define functions and rules using the CLIPS deffunction and defrule constructs.

The CLIPS function “build” will interpret a string as if it has been typed to the interpreter. This may be used to interpret defrule and deffunction messages from other modules, as shown by the rule “interpret-def-commands”. The command “mv-append” is used to compose a list with the desired commands. Build replies TRUE if successful and FALSE if unsuccessful. In either case, the response is sent back to the sender by the function reply.

```
(defrule interpret-def-commands
    (declare (salience 100))
    ?m <- (message ?sender ?command ?body)
    (test (member ?command (mv-append deffunction defrule)))
=>
    (reply ?sender (build ?body))
    (retract ?m)
```

```
)
```

Functions may be defined at initialisation or by messages from other modules. If ?command corresponds to a previously defined function, then ?body may be executed using the CLIPS "eval" command, as shown by the rule "interpret-function-messages". If the message evaluates to a "NIL", then reply does not send a message.

```
(defrule interpret-function-message
    (declare (salience 100))
    ?m <- (message ?sender ?command ?body)
    (test (member ?command (mv-append list-deffunctions))
=>
    (reply ?sender (eval ?body))
    (remove ?m)
)
```

3.3 Automatic Interpretation by Demon Processes

A demon is an automatic procedure which operates on the internal model of each module to detect events. Currently active demon procedures are executed after the update phase of each cycle. Demons are responsible for event detection, and play a critical role in the control of reasoning. Some of the demon procedures, such as motion detection, operate by default, and may be explicitly disabled. Most of the demons, however, are specifically designed for detecting certain types of structures.

The control part of a demon is encoded as rules while the processing part is generally in C. Demons may be invoked by other demons or by commands received from other modules, including from a human supervisor. A demon is instantiated by entering a demon token in working memory. A demon token is simply a list with three elements:

```
(demon  <name> <id>)
```

where <name> is the name of the demon and <id> is a unique identity determined by the function "gensym". Multiple copies of the same demon can be instantiated, each having its own "id". Each demon can create its own state in working memory, indexed by <id>. A demon can be removed by removing the demon token from working memory.

Having a rule interpreter provides makes it possible to have explicit control knowledge for demons and their control logic. It also permits the working memory to be used to create and free working memory for representing demons state. The result is a flexible, easy to use, tool for experiments in control of perception. In the following section we present an example of such control.

4 The Visual Navigation Demonstrator

This section illustrates the use of SAVA III by presenting an overview of the a visual navigation system constructed for the milestone 1 demonstration of VAP-II presented in June 1993. We first provide an overview of the demonstration and the configuration of the SAVA III system which was prepared for the demonstration. We then describe technical details of how vision is used to detect and avoid obstacles and to locate and use landmarks for position estimation. In the final section we describe how rule based control can be used to operate competing visual tasks and react to events.

4.1. Demonstration Overview

Perception serves at least two purposes for the navigation system of a mobile robot [Crowley 85]:

1) Correction of the estimated position as maintained by odometry.

2) Detection and location of obstacles.

Our demonstration uses an interactive graphics tool developed for the MITHRA project to construct a floor-plan "map" of the environment [Causse-Crowley 93]. This map contains walls, furniture, doorways, and visual landmarks (beacons). The tool allows us to specify a number of attributes for each object. In the case of a visual landmark, we can include a symbolic name for its shape, and a three dimension position for a reference point. The interactive specification software also permits us to overlay a network of places and routes on the environment. Using this network we can specify a navigation mission consisting of tasks of the form "go-to <place>" [Crowley 87].

The visuals tasks of obstacle detection and beacon location are contradictory. In order to detect obstacles, the head must look in front of the robot. In order to detect landmarks, the head must look at a wall. Further more, when an obstacle is detected, the system must react to its presence and use vision to steer the robot to a free path. We use a rule based control system to balance these competing activities. This rule base implements the finite state graph shown in figure 6.

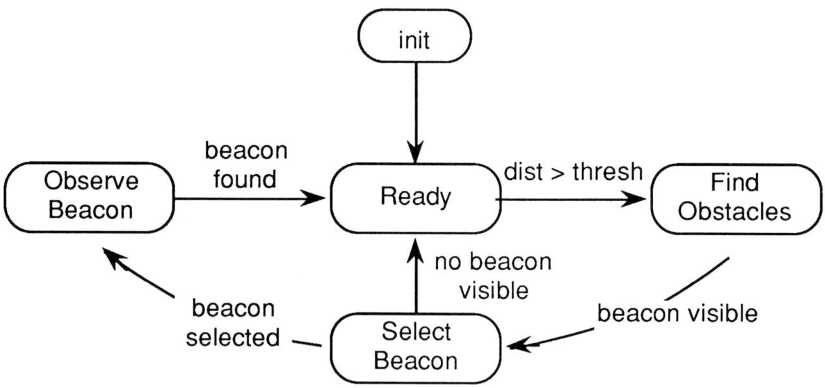

Figure 6 State Transitions Graph for perceptual control of navigation demo

A typical mission involves navigating from a corner of our laboratory to a passage between computer terminals, around a corner and out the door of the laboratory. Once out the door, the vehicle turns around and returns to its starting position.

At the beginning of the mission, the robot verifies that the floor-plan in front of him is free of obstacles. It then locates the nearest beacon and measures the angle. The angle to the beacon provides a one dimensional measurement for updating the estimated position and orientation using an extended Kalman filter. The robot then verifies again that its path is free. Both of these activities occur while the robot is moving.

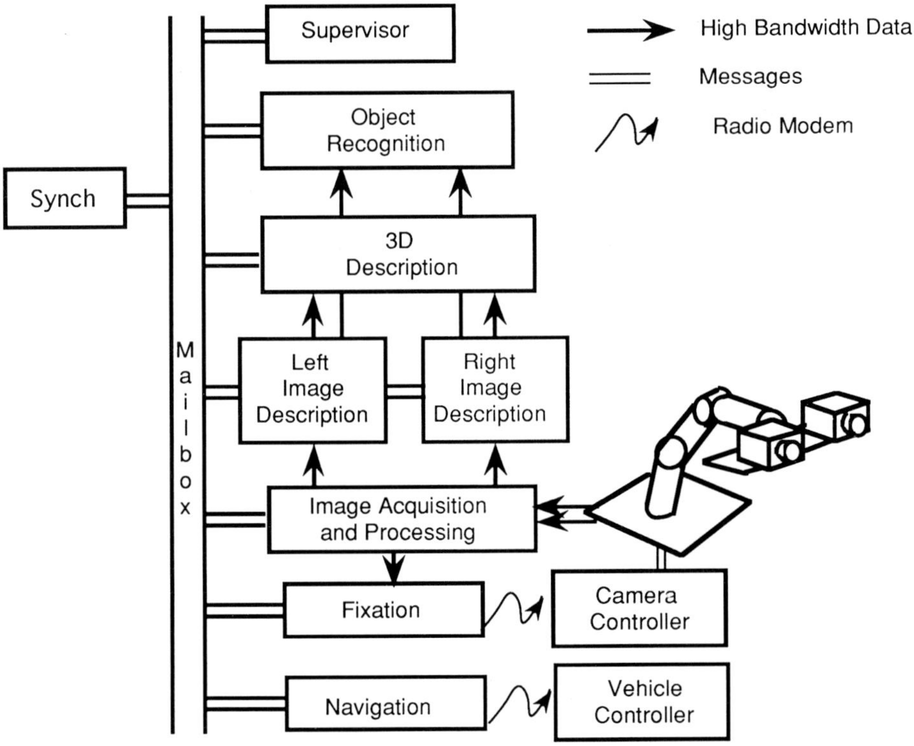

Figure 7 A Distributed Multi-process Vision System.

The structure of the demonstration system is shown in figure 7. The system is composed of processes for

1) Fixation control of the binocular head.
2) Local navigation actions for a mobile robot.
3) Image acquisition and processing.
4) Tracking and grouping a 2-D description of the contents of the image.
5) Computing and maintaining a 3-D description around a fixation point.
6) Recognition of landmarks and object.
7) System Supervisor for coordinating processing of the other system modules.

Fixation Control Unit

The fixation control unit provides a standard interface to the device controller for the VAP/SAVA binocular stereo head. This module maintains a copy of the current state of the fixation point and the component axes for the binocular head. It receives commands in the form of tasks expressed in either device or motor coordinates. Commands are communicated to the binocular head, the robot-arm (neck) or the mobile platform using such device-level control.

The fixation control unit also contains facilities for programming procedural style "perceptual actions". Such perceptual actions are reflex procedures that command the state of the binocular head at either the device level or the motor level based on measurements made from images. Examples of low level perceptual actions include ocular reflexes for servoing aperture, focus and vergence. Other examples include procedures for tracking a moving object.

Image Acquisition and Processing

The image acquisition and processing module handles all image processing requirements for the other modules, thus minimizing the communication requirements. This module is based on two computer cards constructed by the consortium. The first of these, the Pyramid card digitizes synchronised stereo images and immediately computes a 12 level binomial pyramid for the two images. Processing time for each pair of images is 40 ms. The second card extracts edge segments using Gaussian derivatives.

The edge extraction process begins by calculating the horizontal and vertical derivatives within the region of interest. These derivatives are then combined by table look-up to compute the gradient magnitude and orientation. Points which are extrema in magnitude are marked as potential edge points and compared against two thresholds. Hysteresis thresholding is applied so that only regions of edge points containing at least one point above the threshold are considered. Adjacent edge points with a similar orientation are grouped to form line segments. Edge segments are represented by a vector of parameters that includes the mid-point, orientation and half-length.

Three classes of image processing procedures are available in the image processing module

1) Edge Segment Extraction. On command, the module will transfer the pixels within a region of interest to an edge extraction card produced by

the consortium. This card computes the gradient magnitude and orientation and detects pixels which are extrema in gradient magnitude. Detected pixels are grouped in single raster scan to construct edge segments. Gradient magnitudes are compared to two thresholds to provide a hysteresis based thresholding.

2) Edge Chain Extraction: In place of edge segments, another module may request edge chains. Edge points are are computed by the same algorithm as for edge segments. A one pass raster-chaining algorithm is used to construct a list of edge chains within the region of interest. The edge chaining code is computed on a co-processor card.

3) Measures for Ocular Reflexes. In order to avoid communicating images, the measurements on which ocular-motor reflexes are based have been placed in this module. Measures include coarse to fine computation of phase for convergence, and gradient based measurements for aperture and focus.

Image Tracking and Description

An image description is maintained by a tracking process which uses a first order Kalman filter to track edge segments. This tracking process improves the stability of image primitives, permits the system to maintain correspondence of image features over time, and provides an estimate of the position and velocity of image primitives as well as the uncertainty of these estimates. It also permits information about the movement of the head or vehicle to be used to compensate for movements by the robot. A vocabulary of model access and grouping procedures give associative access to the 2D description modules. These procedures are used by a library of "demon" procedures which can be enabled in order to provide data driven interpretation of the image description.

Separate image description modules exist for the right and left cameras. The 2D image descriptions are maintained by a tracking process that uses a first order Kalman filter to track image description primitives. This tracking process improves the stability of image primitives, permits the system to maintain correspondence of image features over time, and provides an estimate of the position and velocity of image primitives as well as the uncertainty of these estimates.

Model access primitives use matching and grouping to interrogate the contents of the token model. A set of demons may be invoked by other modules to interrogate the description after each update using the model access primitives. Access to the 2D model is provided by a large vocabulary of model access and grouping procedures. The four classes of model access procedures are:

Class 1 Procedures for extracting parametric primitives that are close to an ideal geometric entity such as a line or a point. Such primitives are easily implemented and have a complexity that is proportional to the number of primitives in the model.

Class 2 Procedures for extracting parametric primitives that are similar to an ideal "prototype". The prototype is specified an estimated value and a standard deviation for each of the primitive parameters. Similarity is measured by a Mahalanobis Distance. Parameters may be labelled as a "wildcard" by specifying a large (or infinite) standard deviation.

Class 3 Procedures for extracting pairs of primitives that satisfy some geometric relation, such as a junction, an alignment, or a parallelism. The geometric relation is defined as a set of parameters and specified as an estimate and a standard deviation.

Class 4 Procedures for extracting grouping of groupings. If performed by brute force, complex forms such as lines and contours composed of sets of line segments can require an exponential number of computations. By structuring the interrogation as a grouping, we can limit the theoretical complexity to $O(N^3)$ and maintain an average case cost that is nearly $O(N)$.

It is also possible to compose sequences of these grouping procedures, extracting, for example, all the junctions near an ideal line. These procedures may be called by other modules within the system, or they may be invoked by a set of interpretation demons. These demons are placed on an agenda by messages from other modules. After each update cycle the demon agenda is executed.

3D Geometric Scene Description Module

In addition to a description of images, the skeleton system maintains a geometric description of the scene. This geometric description expresses the structure within a region of interest of the scene in terms of 3D parametric primitives. This module assumes that the phase based convergence reflex maintains the cameras converged on an object. Convergence maintains edge segments from a region of interest in the scene in the similar positions in the image. The image description access primitive "FindPrototypeSegment" is used to construct a list of possible matching segments in the left and right image. This list is sorted based on similarity of length, orientation and position. The most likely matches are selected for 3D reconstruction.

Reconstruction requires camera calibration. A novel procedure for dynamic auto-calibration of cameras has been developed. This procedure permits a reference frame for a pair of stereo cameras to be constructed for any scene objects. The projective transformation matrices from object centered coordinates can be obtained by direct observation (no matrix inversion) and can be maintained by a very simply operation. These matrices make it possible to reconstruct the 3D form of objects in an object centered reference frame. As with the image tracking and description module, the geometric description is maintained by a tracking process in order to provide stability and to maintain correspondence over time.

Symbolic Scene Interpretation

The symbolic scene interpretation maintains a symbolic description of the scene in terms of known object categories (or classes) and qualitative relation. This description is built up and maintained by interrogating the contents of the image and scene description modules. The SAVA III symbolic description process was implemented using the CLIPS rule interpreter system. Rules implement a hypothesize and test process which is triggered by demons. Working memory of the production system serves as a blackboard into which recognition procedures can poste their results.

Process Supervisor

The process supervisor maintained a list of places and routes which the system is to travel, as well as a data base of "landmarks" which the system is to find during mission execution. The supervisor plans a navigation which it then executes by sending commands to the other modules. An interesting aspect of the supervisors operation was coordinating between the competing tasks of watching in front of the robot for obstacles and searching for landmarks for position correction. Obstacles must be searched at least once every 50 cm, while landmark detection is required whenever the uncertainty of the estimated position passes a certain threshold. Both operations require command of the camera head. This balancing act was performed by a finite state automata programmed as a set of rules.

Navigation Control

The navigation module controls vehicle actions by sending commands to an on-board vehicle control program. The on-board program, known as the "standard vehicle controller", provides asynchronous independent control of forward translation and rotation. The on-board controller acts like auto-pilot, stabilizing the vehicle and estimating its position. The controller accepts both velocity and displacement commands, and can support a diverse variety of navigation techniques. The controller is capable of responding to commands at any time using a simple serial line protocol. New commands for displacement immediately replace previous commands. This permits visual servoing to be used to pilot the vehicle.

Position and orientation are modelled in the vehicle controlled using Kalman filter to maintain an estimated position and covariance. The control protocol includes a command to correct the estimated position and orientation and their uncertainty from external perception using Kalman filter update. This command has been used to update the estimated position by observing the angle to known objects. The LIFIA standard vehicle controller is described in greater detail in [Crowley-Reignier 93]. The navigation module contains procedures to detect and avoid obstacles, and to locate and use landmarks for updating the vehicle's estimated position.

4.2. Visual Obstacle Detection

The process for obstacle detection is a good example of the implementation of a task-specific vision process within the SAVA system. Our experience with vision for obstacle detection indicates that none of the current techniques for image description are sufficiently robust to work in all cases. Accordingly we have designed vision processes for detecting specific pre-defined classes of obstacles. Our vehicle is equipped with 24 ultrasonic range sensors which function reasonably well for flat walls (with a textured surface) and large objects such as cabinets, but which are very unreliable for the detection of table legs and chair legs. Because table and chair legs are a particular problem, we have defined a vision process specifically for this case.

The 2D description module in SAVA III uses edge segments and grouping of edge segments to describe the contents of the image. The problem with edge segments is that they are often broken at unpredictable points along a contour. Thus we have defined an obstacle detection procedure which detects edges which are not on the floor plane based on the difference of orientation of edges in the right and left image, under a homographic transformation.

Given a pair of stereo cameras, it is very easy to write a 3 x 3 homogeneous coordinate matrix which represents a projective transformation from one image to the other for points within the ground plane. Let ${}^{G}P$, represent a point on the ground plane, and let ${}^{l}P$ and ${}^{r}P$ represent the image of this point in the left and right camera. The 3 x 3 transformation matrix ${}^{l}_{r}\mathbf{D}$ will transform the point ${}^{r}P$ into a point ${}^{l}P'$ in the left image. For the image of all points on the ground plane,

$$ {}^{l}P = {}^{l}_{r}\mathbf{D}\ {}^{r}P. $$

Points above the ground plane will be projected to the left of the corresponding point in the left image, as shown in figure 8. Points below the ground plane (reflections) will be projected to the right. We can use this mapping to detect vertical edge segments for which one of the end points is near the ground plane than the other. Such features include table legs, chairs and door frames, all of which are crucial features for our robot.

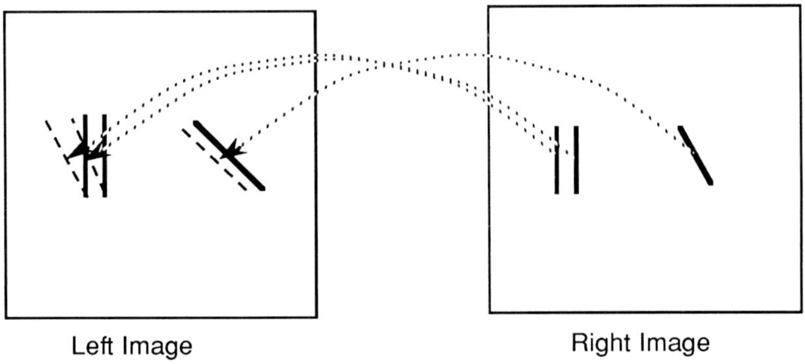

Figure 8 Obstacle detection by mapping edge segments using the homographic transformation defined by the floor plane. Horizontal segments in the floor plane map to segments in the left image with the same orientation. Vertical segments map to a different orientation.

Obstacle detection operates for a given head position and camera vergence angles. With our current 25 mm lenses, at a height of 1.5 meter, at an angle of 45 degrees, the field of view of the floor is a rhomboidal section whose size is very nearly that of a rectanlge composed of four sheets of A4 paper. (or a single sheet of A2 paper). This rhomboidal region constitutes the region of interest in which which visual features constitute potential obstacles. We have defined four arm-head configurations with overlapping fields of view such that the composition is a region of a distance from 1 to 2 meters in front of the robot and roughly 1 meter wide. For each position, a 3x3 homographic transformation is calibrated by imaging a piece of A2 paper decorated with dark squares. Calibration of this matrix is performed by a simple least squares technique [Crowley et al 93].

The visual features which are available are edge segments, expressed both as end-points and as well as in the midpoint, direction and length representation [Crowley et al 92]. Since very short segments have unreliable orientations, we select segments from the right image whose length is longer than a minimum value, typically 6 pixels. The end-points for such segments are projected to the left image to create a list of edge segments from the right image for which we will seek a match.

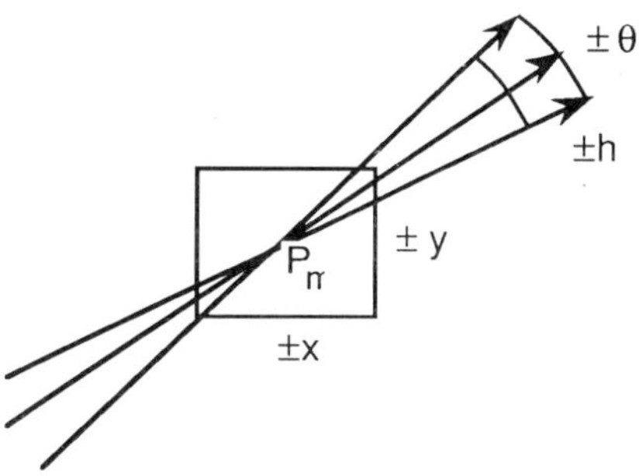

Figure 9 Similarity with respect to a prototype segment.

The SAVA 2D description module contains a vocabulary of procedures for associative access. One such procedure is called Get-Prototype-Segment, illustrated in figure 9. The arguments for this procedures are a vector of parameters $(x_p, y_p, h_p, \theta_p)$, and a set of standard deviations for matching $(\sigma_x, \sigma_y, \sigma_h, \sigma_\theta)$. The procedure Get-Prototype-Segment computes the similarity in orientation, position and length between the prototype segment and each segment in the model. All segments which have a difference less than one standard deviation in orientation, position and length are placed in a list of possible matches. This list is sorted based on the combined Mahalanobis distance, defined by:

$$\text{Dist} = \left(\frac{x_p - x}{\sigma_x}\right)^2 + \left(\frac{y_p - y}{\sigma_y}\right)^2 + \left(\frac{h_p - h}{\sigma_h}\right)^2 + \left(\frac{\theta_p - \theta}{\sigma_\theta}\right)^2$$

In order to use "Get-Prototype-Segment" to detect matching segments, we must furnish estimated uncertainties for the difference in orientation, position and length. The position tolerance is a sum of factors due to the uncertainty of the vergence angle of the head, and the precision of detection of segments. A value of 10 pixels is typical. The length uncertainty is set to the half length of the mapped (prototype) segment. The orientation angle must be large because it is precisely the difference in orientation that signals an obstacle.

The worst case difference in orientation for mapped segments will occur for segments which are perpendicular to the ground plane. By observing such segments, we note that for our camera configuration, such segments have a difference in orientation of about 15 degrees. To add robustness to noise, we set the orientation standard deviation to 20°.

Any segment in the right image for which no candidate match is found in the left image is rejected. For each projected segment for which a match is found, the difference in orientation between the projected segment and the corresponding segment is computed. If this difference in orientation exceeds 10°, then the segment is considered as an obstacle.

When an obstacle segment is detected, its position in the left image is sufficient to compute the necessary turning angle. A bounding rectangle is computed for all obstacle segments. If the center of this rectangle is to the left of the image center, then a right turning angle is computed, otherwise a left turning angle. The turning angle is obtained from the pixel position of right (or left) edge of the rectangle. This pixel value is transformed to degrees using a calibrated ratio of pixel to degrees. The robot is then turned by this amount, and the obstacle test is repeated.

4.3. Visual Localisation

The position state of our standard vehicle controller is the vector $[x, y, \theta]^T$ which gives the vehicles current position and orientation in a global coordinate space. Our standard vehicle controller maintains an estimated of the position vector and its covariance matrix using a Kalman filter fed by odometry. The covariance matrix for this position estimation grows as the vehicle moves. The principle cause of position uncertainty is uncertainty in the orientation, which acts as a lever arm. Uncertainty orientation arises both from from drift while moving and from errors in the estimated center of rotation when turning. The vehicle must use observations of the external environment to keep the position state vector aligned the the real world.

The vehicle controller provides a command for correcting estimated position using an extended Kalman filter. In previous work we have developed and demonstrated a position correction technique based on matching walls observed with a set of 24 ultrasonic range sensors [Crowley 89]. In this demonstration we adapt the vehicle controller to update the position state from the observation of known land-marks in the visual environment. This technique is an adaptation of the technique developed in the thesis of Chenavier [Chenavier-Crowley 92].

A world model is provided to the navigation controller. This model contains a description of walls, obstacles and landmarks. Landmarks have a "type" which symbolically names their shape, a 3D position, and a visibility cone. The navigation controller maintains a list of landmark in the current room, and periodically updates the Cartesian distance to each landmark. Each time the localisation state becomes active, the navigation controller selects a visible landmark. The landmark location is transformed to robot centered coordinates using the estimated position of the vehicle. Fixation is directed to the 3D position of this landmark, using a command to the fixation controller. Perceptual grouping procedures in the 2D description module are used to detect the position of the landmark in the image. Visual servoing is used to bring the landmark reference point to the center of the image. The observed angle to the landmark is then determined from the head controller. The difference between the observed and predicted angle provides a correction vector for an extended Kalman filter.

Landmarks are located using the perceptual grouping procedures in the 2D description module. A demon has been placed in the 2D description module for each class of landmark. Initial experiments have been performed using dark triangles on paper posted on the wall. We permitted the use of artificial land-marks to concentrate on debugging of the basic navigation control processes. We have recently begun using computer screens, which provide easily detected rectangles. There are about 15 within the robot's experimental environment. Triangles and computer screens are detected using the procedure for groupings of groupings. We plan to write landmark detection demons for vertical objects such as pillars, window frames and doors in the near future. Such vertical objects will be detected as a pair of anti-parallel segments with a certain separation.

Updating Position from Beacon Observation

The angle to a known landmark provides a 1-D constraint on position that can be used with an extended Kalman filter to update the vehicle's estimated position and covariance. The key to such an estimate is the model of the observation process, ${}^{Y}_{X}\mathbf{H}$. The vehicle's state vector is the estimation position and orientation.

$$\hat{X}(t) \equiv [\, \hat{x}_r(t),\ \hat{y}_r(t), \hat{\theta}_r(t)]^T$$

This state vector is accompanied by a covariance matrix, $\hat{C}_x(t)$, estimated from an odometric error model. The observation is the relative angle to the landmark. This observation angle is predicted from the position of vehicle and the position of the landmark $\hat{X}_b = (x_b, y_b)$. The predicted vector to the landmark is the difference in position between the robot and the landmark, $(\Delta x(t), \Delta y(t))$

$$\hat{\psi}(t) = {}^{Y}_{X}\mathbf{H} = F(\hat{x}_r(t),\ \hat{y}_r(t), \hat{\theta}_r(t), x_b, y_b) = \mathrm{Tan}^{-1}\left(\frac{\Delta y(t)}{\Delta x(t)}\right)$$

As this observation function is non-linear, we must estimate it with a first derivative taken near the expected value. Thus we replace the observation model with a linear approximation given by the Jacobian of the function $F(\hat{X}(t), \hat{X}_b)$.

$$ {}^{Y}_{X}\mathbf{H} \approx {}^{Y}_{X}\mathbf{J} = \frac{\partial F(\hat{X}(t))}{\partial \hat{X}(t)} $$

The Jacobian is a row vector which contains the derivative of $F(\hat{X}(t))$ with respect to each term in the state vector. Having written the Jacobian approximation to ${}^{Y}_{X}\mathbf{H}$, the rest is classical extended Kalman filter. The Jacobian permits us to estimate the uncertainty of the prediction as a variance based on the vehicle's estimated position.

$$ \sigma^2_{\psi} = {}^{Y}_{X}\mathbf{J}\ \hat{\mathbf{C}}_x(t)\ {}^{Y}_{X}\mathbf{J}^T $$

The observation of a landmark gives an observation angle $\psi(t)$. The difference between the predicted angle and the observed angle gives a correction term $\Delta\hat{\psi}$. The Kalman gain matrix tell how much this correction term contributes to updating each component of the vehicle's state. The vector is computed as a relative waiting of the uncertainty of the position prediction and the observation. If we assume that the observation is perfectly precise, the equation simplifies to :

$$ \mathbf{K}(t) := \mathbf{C}^*_x(t)\ {}^{X}_{Y}\mathbf{J}\ [\ \sigma^2_{\psi}]^{-1} $$

Where ${}^{X}_{Y}\mathbf{J}$ is the derivative of each term in the state vector as a function of the observed direction, $\psi(t)$. A transpose of ${}^{Y}_{X}\mathbf{J}$ is commonly used for this term. $\mathbf{K}(t)$ is a three one vector which tells the contribution to each term of the state vector $\hat{X}(t) \equiv [\ \hat{x}_r(t), \hat{y}_r(t), \hat{\theta}_r(t)]^T$ given by the observation $\Delta\psi(t)$. With this gain matrix we can update the state vector and its uncertainty by :

$$ \hat{X}(t) := X^*(t) + \mathbf{K}(t)\ [\Delta\psi(t)] $$

and

$$ \hat{\mathbf{C}}(t) := \mathbf{C}^*(t) - \mathbf{K}(t)\ {}^{X}_{Y}\mathbf{J}\ \mathbf{C}^*(t) $$

Experiments with an earlier implementation of this process appear in [Chenavier-Crowley 92].

5. Conclusions

According to the principle of "purposiveness" a vision system operates in order to furnish an observation function for some task. In order to enrich our task domain, we have adapted the VAP Skeleton system to serve as the visual component for a mobile robot navigating in an indoor environment. We have then used to the embedded components of the skeleton system to construct visual "competences" which may serve the navigation task. These competences are used as needed by the navigation system. We stress that the visual navigation is not, in itself, the goal of the project. Visual navigation is a task which is sufficiently rich in events to explore the problems of integration and control of an active perception system.

During the last four years, the VAP consortium has constructed a number of demonstrations of continuously operating vision systems. In each of these systems explicit control of sensor motion and processing has permitted the system to operate in real time, with increasingly degrees of robustness. The consortium experience has verified the VAP hypothesis that control of continuously operating process is basic to the design of a general purpose real time vision system.

The construction of these demonstration systems has led us to construct a distributed system composed of a number of cyclic processes. By placing an interpreter at the heart of each of these modules we have obtained a tool which permits experiments in control of perception with a minimum of effort. The fact that the control logic is declarative has greatly enhanced the clarity and simplicity of our experiments. SAVA III has become a tool which has permitted us to demonstrate that a continuously operating active vision system requires control of perception.

Bibliography

[Aloimonos et. al. 1988] Aloimonos, J. Y., I. Weiss, and A. Bandyopadhyay, "Active Vision", International Journal of Computer Vision, Vol. 1, No. 4, Jan. 1988.

[Bajcsy 88] R. Bajcsy, "Active Perception", IEEE Proceedings, Vol 76, No 8, pp. 996-1006, August 1988.

[Ballard 91] D. Ballard, "Animate Vision", Artificial Intelligence, Vol 48, No. 1, pp. 1-27, February 1991.

[Bolles-Horaud 1984] Bolles, R. C. and Horaud, P., "Configuration Understanding in Range Data", Second ISRR, August, 1984.

[Brooks 81] R.A. Brooks, "Symbolic Reasoning among 3-D models and 2-D images", Artificial Intelligence, Vol. 17, pp. 285-348, 1981.

[Brown 90] Brown C. , "Prediction and Cooperation in Gaze Control", Biological Cybernetics 63, 1990.

[Causse-Crowley 93] O. Causse and J. L. Crowley, "A Man Machine Interface for a Mobile Robot", IROS '93, Tokyo, July 1993.

[Chehikian 91] Chehikian A., "A One Pass Edge Exctraction Agorithm" 7th Scandanavian Conference on Image Analysis, Aalborg, 1991.

[Crowley 85] Crowley, J. L., "Navigation for an Intelligent Mobile Robot", IEEE

Journal on Robotics and Automation, 1 (1), March 1985.

[Crowley 87] Crowley, J. L., "Coordination of Action and Perception in a Surveillance Robot", IEEE Expert, Vol 2(4), pp 32-43 Winter 1987. (Also in the 10th I.J.C.A.I., Milan 1987).

[Crowley et. al. 88] Crowley, J. L., P. Stelmaszyk and C. Discours, "Measuring Image Flow by Tracking Edge-Lines", Second ICCV, Tarpan Springs, Fla. 1988.

[Crowley 89] Crowley, J. L., "World Modeling and Position Estimation for a Mobile Robot Using Ultrasonic Ranging", 1989 IEEE Conference on Robotics and Automation, Scottsdale AZ, May 1989.

[Crowley et. al. 90] Crowley, J. L., P. Bobet and K. Sarachik, "Dynamic World Modeling using Vertical Line Stereo", First European Conference on Computer Vision, (ECCV-1) Antibes, France, 1990.

[Crowley 91] Crowley, J. L. "Towards Continuously Operating Integrated Vision Systems for Robotics Applications", SCIA-91, Seventh Scandinavian Conference on Image Analysis, Aalborg, August 91.

[Crowley et. al. 92] J. L. Crowley, P. Stelmaszyk, T. Skordas and P. Puget, "Measurement and Integration of 3-D Structures By Tracking Edge Lines", International Journal of Computer Vision, Vol 8, No. 2, July 1992.

[Crowley-Chenavier] F. Chenavier and J. L. Crowley, "Position Estimation for a Mobile Robot Using Vision and Odometry", 1992 IEEE Conference on Robotics and Automation, Nice, May , 1992.

[Crowley et al. 93] J. L. Crowley, P. Bobet et C. Schmid , "Auto-Calibration of Cameras by Direct Observation of Objects", Image and Vision Computing, Vol 11, no. 2, March 1993.

[Crowley-Christensen 93] Vision as Process, Results from Project ESPRIT BR 3038, Springer Verlag Basic Research Series, to appear 1993.

[Crowley-Demazeau 93] J. L. Crowley et Y. Demazeau, "Principles and Techniques for Sensor Data Fusion", Signal Processing, March 1993.

[Crowley-Reignier 93] J. L. Crowley et Patrick Reignier, "Asynchronous Control of Rotation and Translation for a Robot Vehicle", Robotics and Autonmous Systems, Vol 10, No. 1, January 1993.

[Draper et al. 89] Draper B. A., R. T. Collins, J.Brolio, A. R. Hansen, and E. M. Riseman, "The Schema System", International Journal of Computer Vision, Kluwer, 2(3), Jan 1989.

[Eklundh 92] Eklundh, J. O. and K.Pahlavan, Head, "Eye and Head-Eye System", SPIE Applications of AI X: Machine Vision and Robotics, Orlando, Fla. April 92.

[Granum-Christensen 1990] E. Granum & H.I. Christensen, "Dynamic Robot Vision, In: Traditional and Non-traditional Robotics Sensors", T. Henderson (Ed.), NATO ASI

Series in Computer Science, Springer Verlag, 1990.

[Hanson-Riseman 1978] Hanson, A.R. & Riseman, E.M., "VISIONS: A Computer Vision System for Interpreting Scenes", in Computer Vision Systems, A.R. Hanson & E.M. Riseman, Academic Press, New York, N.Y., pp. 303-334, 1978.

[Hanson-Riseman 1987] A.R. Hanson & E. Riseman, "A Knowledge Based Approach to Vision. In: Vision, Brain and Co-operative Computation", (Eds.) M.A. Arbib & A.R. Hanson, MIT Press, Cambridge, Mass, 1987.

[Huang 83] Huang T. H., Image Sequence Processing and Dynamic Scene Analysis, Springer Verlag, Berlin, 1983.

[Kalman 60] Kalman, R. E. "A New Approach to Linear Filtering and Prediction Problems", Transactions of the ASME, Series D. J. Basic Eng., Vol 82, 1960.

[Kalman 61] Kalman, R. E. and R. S. Bucy, "New Results in LInear Filtering and Prediction Theory", Transaction of the ASME, Series D. J. Basic Eng., Vol 83, 1961.

[Krotkov 90] Krotkow, E., Henriksen, K. and Kories, R., "Stereo Ranging from Verging Cameras", IEEE Trans on PAMI, Vol 12, No. 12, pp. 1200-1205, December 1990.

[Marr 1982] Marr, D., Vision, W. H. Freeman, San Francisco, 1982.

[Rimey-Brown 1991] R.D. Rimey & C. Brown, "Controlling Eye Movements with Hidden Markov Models", Intl Journal on Computer Vision, April 1991.

[Tsotsos 87] J.K. Tsotsos, "Representational Axes and Temporal Co-operative Processes, In: Vision", Brain and Co-operative Computation, (Eds.) M.A. Arbib & A.R. Hanson, MIT Press, Cambridge, Mass, pp. 361-418, 1987.

A MODULAR CONTROL ARCHITECTURE FOR

REAL-TIME SYNCHRONOUS AND ASYNCHRONOUS SYSTEMS

PHILLIP L. BUTLER
Robotics and Process Systems Division
Oak Ridge National Laboratory
Oak Ridge, TN 37831-6304 USA

JUDSON P. JONES
Engineering Physics and Mathematics Division
Oak Ridge National Laboratory
Oak Ridge, TN 37831-6364 USA

ABSTRACT

This paper describes a control architecture for real-time control of complex robotic systems. The Modular Integrated Control Architecture (MICA), which is actually two complementary control systems, recognizes and exploits the differences between asynchronous and synchronous control. The asynchronous control system simulates shared memory on a heterogeneous network. For control information, a portable event scheme is utilized. This scheme provides consistent interprocess coordination among multiple tasks on any number of distributed systems. The machines in the network can vary with respect to their native operating systems and the internal representation of numbers they employ. The synchronous control system is needed for tight real-time control over complex electromechanical systems such as robot manipulators and utilizes multiple processors at a specified rate. Both the synchronous and asynchronous portions of MICA have been developed to be extremely modular. MICA presents a simple programming model to code developers and also considers the needs of system integrators and maintainers.MICA has been used successfully in a complex robotics project involving a mobile 7-degree-of-freedom manipulator in a heterogeneous network with a body of software totalling over 100,000 lines of code.

1. Introduction

Robots differ from other intelligent systems in that they interact with the world by mechanical means. Robotic system architectures must simultaneously incorporate manipulation control, multiple sensor systems, operator interfaces, navigation control, symbolic reasoning, and safety systems, to name but a few. They must address the requirements of real-time systems that interact directly with sensors and effectors while simultaneously meeting the needs of higher level reasoning processes that do not interact with the world in a direct sense and cannot meet hard real-time deadlines. Consequently, both the hardware and the software architectures of robotic systems tend to be complex.

From the software point of view, system builders have typically relied upon some kind of centralized structure and a variety of communications schemes to coordinate the activities of various parts of the system. Examples include the NASREM[1] architecture, which incorporates a central memory, and the Task Control Architecture[11], which incorporates a central controller which communicates with remote tasks by message passing. The Ground Surveillance Robot[7] simulates a centralized blackboard on a distributed system where the processors communicate by exchanging messages according to a fixed protocol. Although a centralized resource of any kind will eventually constitute a performance bottleneck if it is physically instantiated, assuming that a central memory or central controller simplifies the software design immeasurably (see Brooks[2] for a noteworthy counterexample).

From the hardware point of view, system builders have been able to take advantage of the physical proximity of their processors to implement common-bus multiprocessing schemes (e.g. CONDOR[8]). Tight coupling between processors permits fast interprocess communications schemes to be implemented on the basis of shared memory in a global address space. Tight coupling also permits real-time multiprocessing because the existence (or the simulation) of a common clock eases interprocessor synchronization. Again, contention for the common resource may constitute a potential bottleneck, but the existence of a common bus, a common clock, and a global address space greatly simplifies system integration.

However, tight coupling may not always be possible, for instance, when a mobile robot must communicate with a remote supervisor or when multiple mobile robots must cooperate to accomplish a single task. It is also desirable in some instances to be able integrate a heterogeneous collection of computers because one may be able to perform a calculation more effectively than another. For example, an operator's console may contain special-purpose graphics hardware and run an operating system different from that of a robot. In these instances, the assumption of a centralized resource, either hardware or software, becomes difficult to maintain.

Nevertheless, the conceptual simplicity of shared memory makes it an extremely attractive programming model, especially in large projects where coordinating the activities of a large number of people can itself become a significant source of complexity. In this paper, we describe the Modular Integrated Control Architecture (MICA), which provides separate mechanisms for both synchronous (hard real-time) and asynchronous (soft real-time) systems and defines an interface between them. MICA integrates tightly coupled real-time multiprocessor systems with a loosely coupled asynchronous heterogeneous network. MICA hides this complexity behind a simple programming model by simulating a shared memory over the heterogeneous network[9] thus easing the conceptual burden on software architects and the implementation burden on system developers. MICA also automatically generates intersystem translation software, thus avoiding a potentially costly maintenance problem in heterogeneous systems.

We begin by discussing the architectural requirements for robotic systems, focusing on the interprocess communication and synchronization requirements of sensor-based robotic systems. We then illustrate a model of a distributed multiprocessor system and present a communications system that eliminates much of the complexity. We then discuss the details of the integration of synchronous real-time subsystems. We attempt to explain the motivations behind many of our design choices by examining their impact on the system development process. We conclude by presenting the implementation of these concepts on a particular robotic system.

2. Architectural Requirements

We classify processes in a robotic system into two types: asynchronous and synchronous. Asynchronous processes may require interprocess synchronization or may be free-running processes that perform some sort of computation without interprocess synchronization. In either case, they do not have strict execution deadlines. Synchronized processes are coordinated by a common real-time clock. Because programming hard real-time systems is generally more difficult, we would like to construct systems composed mostly of asynchronous processes.

MICA is a two-part software architecture, as shown in Figure 1. Processes are divided into an asynchronous group and a synchronous group. Within the asynchronous group, processes communicate with each other either by placing data on a "blackboard", or by sending "events" (messages having a stereotyped format), or both. Any process within the asynchronous group may communicate with any other. The synchronous group is composed of one or more "clusters", processes synchronized by a common clock. Synchronous processes can communicate with each other by posting data on a common block local to the cluster. However, synchronous processes can communicate synchronously only with other processes in the same cluster. If communication with an asynchronous process or another cluster is desired, an asynchronous interface is used.

To demonstrate how these kinds of processes arise naturally in practice, we briefly examine components from a proof-of-concept demonstration for autonomous radiation surveillance of 55-gallon drums[12]. This scenario was chosen because there are a large number ($>10^6$) of such drums in the USA and elsewhere storing low-level radioactive waste. These drums must be monitored essentially continuously to guard against leakage. Currently this monitoring is done manually, is labor intensive and hazardous, and is thus a good candidate for automation.

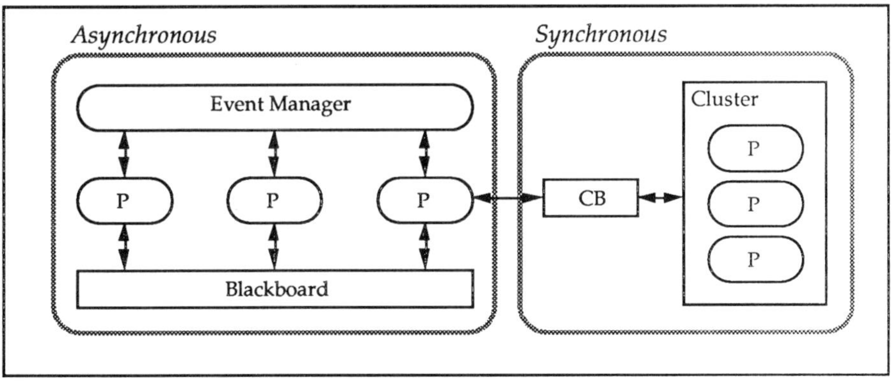

Figure 1.□MICA separates complex systems into a combination of asynchronous processes and synchronous processes. Asynchronous processes utilize a blackboard for data and an event manager to handle control information. Synchronous processes are contained within a "cluster" that utilizes a single common block for data storage and synchronous interprocess communication. A single asynchronous process is the interface between a cluster and the rest of the asynchronous system.

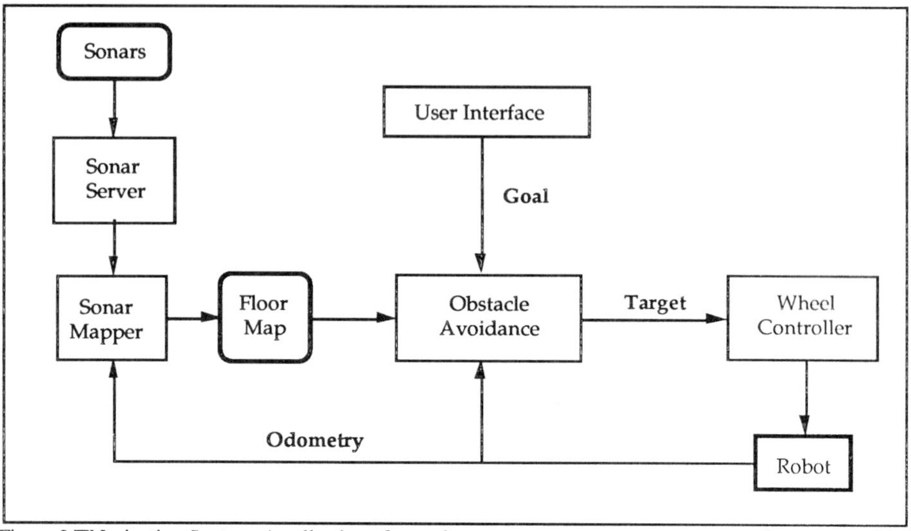

Figure 2. Navigation System. A collection of asynchronous processes cooperate to control a sensor-guided robotic navigation system. Data acquired through sonars are used to produce a floor map. An obstacle avoidance routine inspects this map and sends a command to a wheel controller to move the robot to a goal specified by the user.

The task calls for a mobile robot to navigate autonomously through a collection of a priori unknown obstacles until it reaches a certain position. Once there, it uses a range imaging system to identify all of the 55-gallon drums in its field of view. It then selects a drum and moves to a position relative to the drum that is kinematically acceptable for radiation monitoring. It uses a radiation sensor mounted on a manipulator to determine whether radiation is leaking from the drum (beta radiation levels significantly higher than background). If the drum is leaking, an operator is so informed.

The requirement for interprocess synchronization is typified by the sequence of events required to take the robot from the position where it acquires a range image to the 55-gallon drum. The robot it acquires a range image, analyzes the range image, and chooses a drum. Then it determines an appropriate position relative to the drum and travels to that position. Each of these component tasks is most effectively implemented as an independent process that can be independently tested. The exact time required to execute each component task is not critical, but each task must be accomplished in sequence. Consequently, the tasks require interprocess synchronization.

The requirement for free-running, unsynchronized processes is exemplified by sensor-based navigation. Figure 2 illustrates a simple map-oriented obstacle avoidance system based on ultrasonic data. A "sonar_server" process continuously actuates sonars mounted on a ring encircling the robot chassis. The robot's position in the world coordinate system is stored along with the range returned from each sonar reading. A "sonar_mapper" process runs a ray-tracing algorithm to produce a bit-map of the robot's immediate neighborhood. An obstacle avoidance process examines this map, the robot's current position, and the goal position to produce a target for the low-level "wheel_controller." Each of these processes consumes and produces data essentially continuously, using the latest available data from its input processes. Synchronization is not necessary.

The requirement for real-time subsystems is exemplified by the process of scanning the drum for radioactivity. In this case, we must coordinate a relatively large number of manipulator joints and perform control calculations based on sampled data from optical encoders on the motors. Precise control requires a real-time deterministic mechanism, because of the servo control loops and digital filtering techniques used for manipulator control. Sensor input, servo controllers, and actuator drive outputs must be synchronized with a common real-time clock. This synchronization requirement is typical of systems that acquire data and/or control actuators at a specific rate.

3. Asynchronous System Control

The asynchronous portion of MICA allows asynchronous processes to communicate control and data by using a blackboard for the communication of data and events for interprocess synchronization. MICA runs on a heterogeneous distributed-memory network

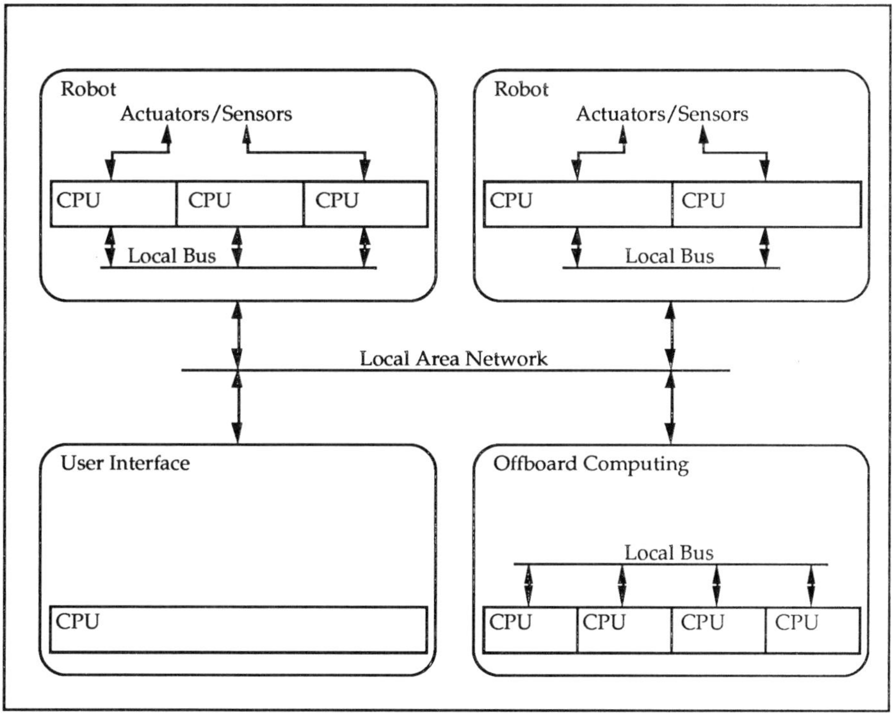

Figure 3.□A typical integrated robotic system consists of one or more robot subsystems, a user interface, and possibly some off-board computing. Each subsystem would contain its own local bus with a single link to a local area network.

of workstations by simulating shared memory (see Nitzbergg and Lo[9] for a recent review of this topic). The various machines on the network can be heterogeneous with respect to the native operating system and the native internal representation of numbers used on each machine. In brief, this is accomplished by replicating the contents of the blackboard into each separate address space in the system. Messages are created from parts of the blackboard and broadcast to each address space in the network when the contents of the part are changed.

A hardware configuration typical of advanced robotics systems is illustrated in Figure 3. In this illustration, two robots each containing a common bus multiprocessor are linked via a local area network (LAN) to a workstation running a user interface and another multiprocessor system performing off-board computation.

Figure 4 illustrates how MICA realizes the software model of Figure 1 on such a system. A replica of the entire blackboard is resident in the memory of each machine.

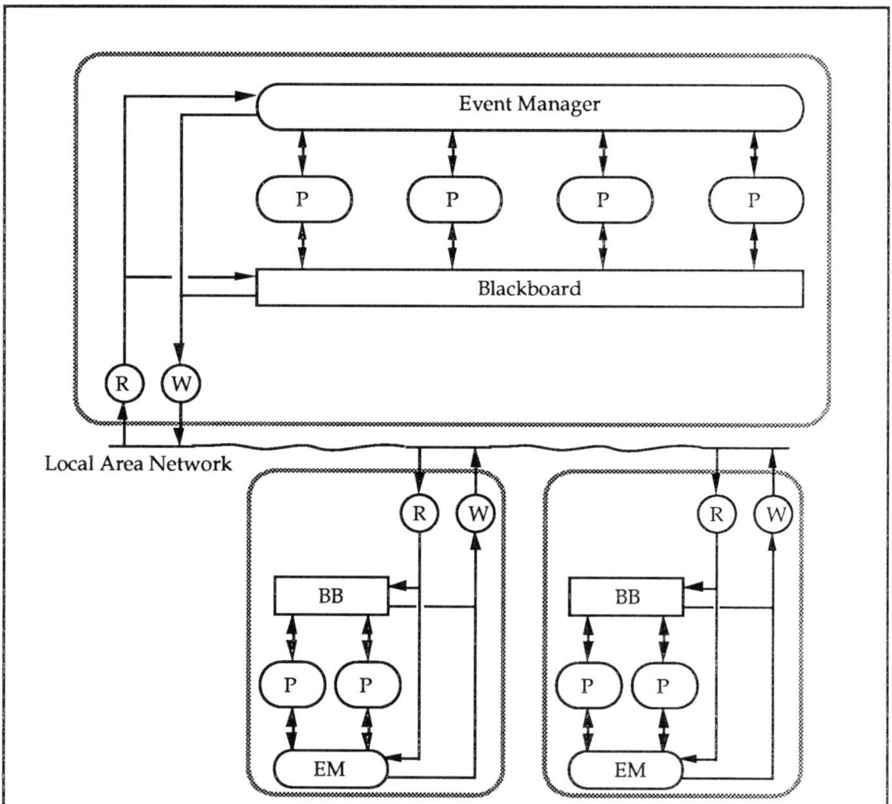

Figure 4.□Distributed systems are composed by replicating and distributing the shared memory though which asynchronous processes communicate. A reader (R) and a writer (W) process on each system handle communication across the local area network.

Asynchronous processes (P), communicate with each other through this blackboard (one or more of these processes may be an interface to a synchronous cluster). Whenever one of these processes updates data on the blackboard, an interrupt awakens a sleeping write process (W), which writes data into the communications network. Data incoming from the network interrupt sleeping read processes (R) on other machines, which store the data in the local copy of the blackboard. An event manager process (EM) also runs on each machine. When an event arrives for a process, it is placed in an input queue specific to that process. If the process is not running on the machine where the event originated, the event manager broadcasts the events to all other machines in the network.

3.1 Blackboard Structure

MICA uses a modified blackboard metaphor[6] for interprocess and intersystem communication. In this modification, all data to be shared between processes, regardless of which processes produce or consume them, are posted on a single large blackboard covered with "Post-Its". Related data are collected together onto individual Post-Its. The Post-Its are concurrent read/exclusive write. Any process can read the data on any Post-It, but only one process can write to a Post-It.

The blackboard has a name, "BB" for example, as do all of the Post-Its (e.g., "odometry," "range_picture," etc.), as do the individual variables on each Post-It. To refer to any variable on the blackboard, we use a syntax that specifies the name of the blackboard, the name of the Post-It, and the name of the variable. For example, to refer to the current orientation of the robot in the world coordinate system, we use the syntax "BB->odometry->orientation".

The blackboard name is actually a pointer to a list of the addresses of all the Post-Its. The Post-Its themselves are kept together in a single large common store. A small set of functions controls access to the data on the blackboard, as shown in Figure 5. One of these

mem_attach()	returns a pointer to the blackboard.
get_read_ok(Post-It)	increments a counting semaphore which records that the data on a particular Post-It are being read.
read_done(Post-It)	decrements the read semaphore.
get_write_ok(Post-It)	waits until the read semaphore is clear, then sets a write semaphore, blocking further reads. The calling process is then assured that it can change the data without corrupting a read.
write_done(Post-It)	clears the write semaphore. As a "side-effect" it sends an interrupt to a "communicate" process which manages intersystem communications. The communicate process broadcasts the contents of the updated Post-It to all other address spaces in the network.

Figure 5.□A set of C language functions are used to coordinate blackboard accesses. Actual reading and writing of the blackboard by processes is through the use of conventional C structure access.

functions is an initialization and the others implement access control. At any given time, any number of processes can read the data on a Post-It, but when a process wishes to write to a Post-It, it is blocked until all the readers finish. Potential readers are blocked until the write is complete.

Coherency of the entire blackboard is not maintained over the entire distributed network. If it were, it would be necessary for a lock on a particular Post-It to be obtained system-wide before a process could update the data on that Post-It. It would then be necessary to verify that the new information had been communicated to all systems before the lock could be released. This approach would require an excessive amount of communication overhead for each update.

Instead, the access control functions act only locally. Processes are blocked from reading Post-Its while they are being updated, but only when they are being updated on the local system. After the update occurs, the new data are communicated to other systems. Reading processes on these destination systems are only blocked while the Post-It is actually changing. If global coherency is required, it may be explicitly implemented on a case-by-case basis by using an event protocol (see below) between the relevant processes. Thus, we have chosen to make the system fast by default, while providing a path by which a slower but safer policy could be adopted.

3.2 Heterogeneous Systems.

The principal problem with heterogeneous systems lies in the different internal representation of numbers used by different machines. The disastrous consequence of this inconsistency is that the actions associated with communicating a message between machines depends upon the contents of the message. For example, to communicate a floating-point variable from one machine to another, it is not sufficient to know only the number of bytes it occupies; one must translate the datum from the format of the source machine into the format of the destination machine.

We address this problem by using a simple compiler to automatically generate translation software given a description of the Post-It data formats. That is, we imagine a machine that has an instruction set consisting of atomic instructions for translating variables from the internal representation of any particular machine into and out of a language common to all machines. Our compiler inspects a file that describes the Post-Its on the blackboard and generates code in this instruction set, which translates the Post-Its into and out of this common representation. The compiler recognizes all of the native storage classes defined in "C" and composite data structures consisting of structures and arrays, nested arbitrarily deeply. Pointers cannot be stored on the blackboard, because an address on one machine is meaningless on another.

The compiler generates a linked-list of linked-lists describing the structure of the data on the blackboard. The outermost list contains one link for each Post-It definition. The innermost list contains a link for each variable in the the Post-It definition. This list of lists is traversed several times to create source code for programs for allocating memory, initializing semaphores, calculating addresses, creating diagnostic tools, and a few other functions; principle among them is translation.

The main advantage of using automatic code generation is in the development, integration, and maintenance phases of a project. In particular, if the detailed design is in error for one reason or another (e.g., perhaps the designer forgot a variable) it is possible to fix all of the communications code in the system simply by running the code generator. The code produced is guaranteed to be correct the first time. Automatic code generation eliminates manual maintenance for a large body of what would otherwise be very expensive code.

3.3 Events

Data on the blackboard are regarded as continuous in time; therefore, MICA utilizes an event mechanism to pass control information from one process to another. An event is a message having a stereotyped format that transmits a small quantity of control information between processes. Event messages provide a simple transparent method to pass control information within a single processor, among processors in a backplane, and among processors in a LAN. Events can be used in concert with the blackboard so that the communication of data can be coordinated with the communication of control information.

The concept of the event system is similar to the event systems typically used in graphical user interfaces such as the Apple Macintosh. In the Macintosh, events are defined to correspond to user input such as mouse clicks and key presses and also to system control messages such as window update events. This organization allows programs to be modeless, responding to messages delivered by the system to the single Macintosh process. The MICA event system takes the event concept one step further by allowing multiple processes to control one another by using events.

3.4 Event record

Events contain the information shown in Figure 6. An event code typically contains the information necessary for action. Some event codes such as NULL_EVENT and QUIT_EVENT are common to all processes. Other event codes are specific to a particular process. For example, an event code may exist to command the sonar process to start a

```
typedef  unsigned short event_code;      /* event code */
typedef  unsigned short event_type;      /* event process id */
typedef  unsigned short event_priority;  /* event priority */

struct event_record
{
  event_code   code;             /* event code */
  event_type   source;           /* source process type */
  event_type   dest;             /* destination process type */
  event_priority priority;       /* event priority */
  unsigned long extra[14];       /* extra data related to event */
};
```

Figure 6.□Processes are synchronized by using events -- messages having a stereotyped format. This C language structure is used by processes to communicate control and interprocess synchronization messages.

register_event(process_name)	identifies a process itself to the event manager.
resign_event()	notifies the event manager upon quitting.
get_next_event(event_record, mode)	returns an event code, and optionally, its associated event record. The calling process can decide whether or not to suspend execution until an event is received by specifying blocking or non-blocking in the mode argument. If no events are available, a non-blocking read returns a NULL_EVENT.
post_event(code, destination, priority, event_record)	sends an event code to any process, including itself. If an event does not use the optional data field, it only has to pass a null pointer as the address of the event record.
event_available(event_record)	returns true if an event is available to be received. A process may need to determine whether an event is pending without actually dequeueing it. The process also can pass the address of an empty event record to get a copy of the pending event and can determine whether it should process the event or defer it until a later time.

Figure 7.□A set of C language functions used to coordinate control between processes.

sonar scan. An event source and destination, included in the event record, record the sending and receiving processes. This information is useful for an event-tracing process that monitors all events communicated in a distributed system. Events are prioritized so that high-priority events such as emergency shutdown supersede any lower priority events. Priorities also can be used to set the sequence of operations for a complex set of event messages. An optional data field can be used for auxiliary information such as desired arm positions, sonar enable masks, and so forth.

A small set of functions is used by developers to handle events as shown in Figure 7. Two of these functions (get_next_event and post_event) are for receiving and sending events. However, like data on the blackboard, events are transmitted across system boundaries. Therefore, the event manager running on each system needs to be kept informed of which processes are running locally on that system. Thus, two other functions (register_event and resign_event) inform the event manager that a process is starting and stopping. Intraprocessor, interprocessor, and intersystem event communication is transparent to both the process sending an event and the process receiving an event.

3. 5 Event Manager

An event manager is used to control the delivery of events from one process to another. The event manager maintains an array of priority queues, one for each process, to hold events until they are requested. The event manager receives an event posted by an individual process and places it into the appropriate priority queue, prioritizing as it is placed into the queue. When a process requests an event, the event manager pulls the top item from that process's event queue and returns it to the process.

An event monitor tool is used to monitor the delivery and execution of events within the asynchronous system. The event monitor assists in the initial development of interprocess event communications by providing an event trace of all events between asynchronous processes. This trace mechanism is used in tracking down problems with interprocess event communications and handshake. In addition, the event monitor trace capability is useful in maintaining code over the long term by helping verify proper event control sequences.

```
struct event_record my_event;
struct blackboard * BB;

main()
{
  int   i, done = FALSE, scan_flag = FALSE;
  double distance;
  event_code event;

  BB = mem_attach();
  register_event(SONAR);
  while (!done)
  {
    event = get_next_event(&my_event, BACKGROUND);
    switch(event)
    {
      case QUIT_EVENT:       /**** Common events ****/
        done = TRUE;
        break;
      case SONAR_SCAN_ON:    /**** Process specific events ****/
        scan_flag = TRUE;
        break;
      case SONAR_SCAN_OFF:
        scan_flag = FALSE;
        break;
    }
    if(scan_flag)
    {
      get_write_ok(BB->sonar);
      for (i=0; i<N_SONARS; i++)
      {
        distance = get_sonar(i);
        BB->sonar->distance[i] = distance;
      }
      write_done(BB->sonar);
    }
  }
  resign_event();
}
```

Figure 8.□Most programs utilizing MICA are written in the form of an event loop. In this example, the program responsible for actuating the sonar ensemble can be enabled or disabled with events.

3.6 *A simple example*

One benefit of using events is that of consistent process implementation. Event processes are written in the form of an event loop, as shown in Figure 8. The event loop receives an event from the event manager and then acts on it. This modeless design allows any event to be processed properly whenever it comes in. Figure 8 also shows an example of blackboard access. Writing to the blackboard is performed in a short section of code, because readers will be blocked until the write is complete.

4. Synchronous System Control

The synchronous portion of MICA provides more precise control with respect to time than does the asynchronous portion. This type of control is needed in many real-time hardware control schemes but can be applied whenever appropriate. A synchronous system can be executed on one or more processors, but they must be driven by a common clock. Because the timing constraints are rigid, synchronous systems require tightly coupled processors if more than one processor is used. The synchronous portion of MICA is modeled after previous work on the NASA Laboratory Telerobotic Manipulator[10] (LTM). LTM uses multiple processors to execute the real-time control algorithms necessary for precise control of dual master/slave manipulator arms. The synchronous portion of MICA is based on the concepts used in LTM but adds increased modularity and extendibility.

The need for a deterministic, time-critical system is based on the needs of modern PID (proportional-integral-derivative) algorithms that are used in most servo controllers. The integral and derivative terms become very complicated if a precise time step is not maintained. Furthermore, such an algorithm would require precise knowledge of the exact time increments between control loops.

Several practical considerations also guided the design of MICA. First, because the synchronous code would normally be controlling complex hardware systems, it is necessary to provide a simulation capability. This capability allows us to simulate new algorithms, gaining confidence in them before application to the actual hardware system. Second, safety code must be appropriately integrated for the system to operate in a controlled manner. Automatic system shutdown should occur if an erroneous motion command is given. In a modular system, safety code can be separated from the control algorithms and tested separately. In this manner, we can test new algorithms with minimal concern to the safety aspects of the system. This also has the advantage that critical safety systems cannot be bypassed. Finally, as system complexity grows, new processors sometimes must be added to bear the burden of increased processing loads. By adding processors, the existing part of a synchronous system should be minimally affected.

4.1 *Synchronous hierarchy*

MICA organizes a synchronous system into a hierarchy, as shown in Figure 9.

Cluster refers to an entire synchronous system. This distinction is necessary because an arm-control cluster may not necessarily be synchronized to a wheel-control cluster. Control code that must operate synchronously is contained within the cluster. Every cluster has an asynchronous interface to allow other processes to communicate with the cluster. This interface is able to interact with both the asynchronous system and the synchronous control loops within the cluster.

Modules are functional entities within a cluster. For example, there may be an input/output (I/O) module and a path-planner module. The collection of modules defines the cluster.

Packages are alternative instances of modules. Installing different packages into a cluster typically changes the characteristics of the entire cluster. For example, a simulation package may be chosen for an input/output module instead of an actual drive package to perform simulations.

Routines perform the desired function of the package. One or more routines can exist in a single package. Several reasons exist for having multiple routines in a package. One is to allow a single package to have code executing over several processors. Another is to allow two different routines to execute in different orders on a single processor. For example, the actual I/O package may have a routine to do input. Another routine may then execute at a later time to perform safety checks and output to the motor drives. It is important to note that once the routines for a package are defined, a package can then be treated as a single entity. Therefore, a package can be installed and removed without having to define explicitly which routines are affected.

4.2 Synchronization mechanisms

A simple naming convention is used for the processors needed for a synchronous systems. A control processor (CP) is used to be the asynchronous interface to the synchronous system. This processor has a conventional, event-driven interface program that also has the ability to control the synchronous system. The CP can also contain

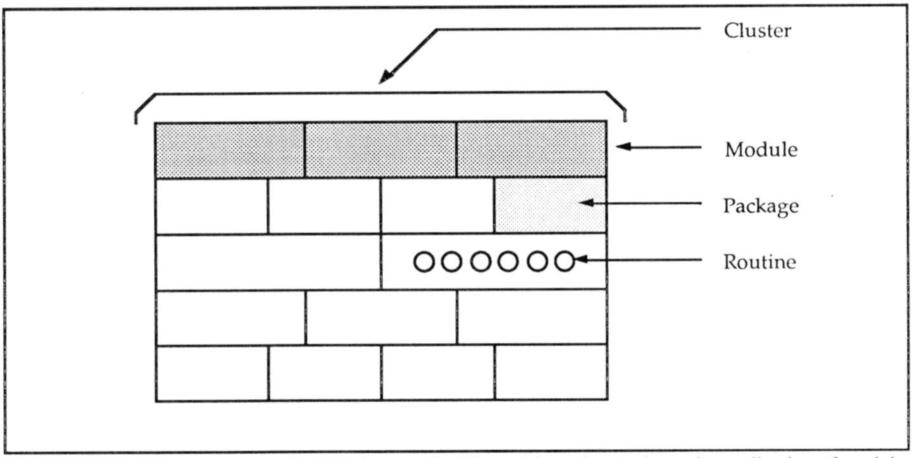

Figure 9.□The synchronous part of MICA is organized into a hierarchy. A cluster is a collection of modules synchronized by a common clock. Modules are composed of one or more packages, that contain the actual code routines. Modules can be changed out to change the function of the cluster.

unrelated asynchronous programs for other parts of the integrated system. The CP can also contain a synchronization interrupt routine that is used to control the synchronous processors, but this is not absolutely necessary.

One or more loop processors (LPs) execute the synchronous control loop. In our initial implementations, the system uses two LPs to execute all control code. The real-time control loops must have the highest priority on the LPs and not be interrupted by asynchronous processes.

The CP contains a synchronization interrupt routine that is used to set sync flags within the multiple LPs as shown in Figure 10. The LP execution code waits in a loop until the sync flag is set. Once the sync flag is set, the LP checks for changes in state and starts sequentially executing a number of "chores." Since a system designer must be able to precisely define the order of execution of routines within each LP, routines are assigned to a chore list to define the sequence of execution for the LP.

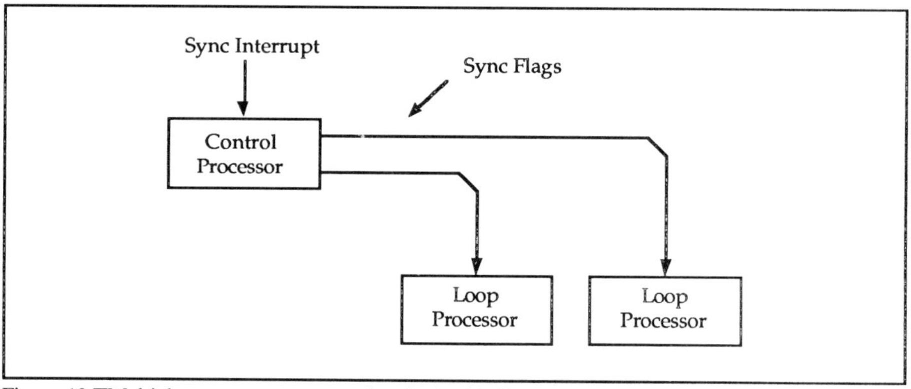

Figure 10.□Multiple processors are synchronized with flags, driven by a clock interrupt. In a typical system, the synchronization interrupt comes from a short interrupt service routine on the control processor but may also come from an external source, thus synchronizing the control loops to an external system.

The chore list is an ordered list of routine function vectors to execute for a particular control loop. A function vector is an indirect pointer to a C function. This C construct allows any function to be executed simply by placing its address in the function vector. The selection of the actual function is deferred until execution time and can be changed quickly. Therefore, different routines can be installed and removed quickly from the chore list. During setup and initialization, the CP assigns different routines to the chore list of the individual LPs. In the future, a chore manager could be created to dynamically assign routines to the LPs across each loop to evenly distribute the processing load.

Once the LP has finished with its chores, it clears its sync flag and starts the control loop again, waiting for the next sync flag. Because the sync flag is cleared after all chores have been completed, the CP can check the sync flag to make sure it is cleared before it sets the sync flag again. If the sync flag is still set, the LP has not finished the loop and a slow loop counter is incremented. In some cases, slow loops indicate a serious problem, causing a system shutdown. In other cases, an occasional slow loop is acceptable, especially when an exception condition occurs that requires more processor time. Many slow loops indicate improper chore assignment among the multiple LPs.

The use of dedicated CPUs is not required for MICA but is an implementation convenience. If other processes were allowed, the synchronous operation probably would be entirely part of a high-priority interrupt routine instead of using the sync-flag method discussed above. In this case, the LP would not execute on a dedicated processor but would be an interrupt routine. Other than this deviation from the terminology, all other synchronous constructs would remain intact. Additionally, a hardware-interrupt controller could be used to synchronize the LPs.

Though routines must be synchronized by MICA, there is flexibility as to the rate of synchronization. An "LP sequence factor" can be applied to the sync interrupt to determine the rate of synchronization for each LP. This allows time-consuming routines to execute at a slower rate while maintaining precise synchronization. In addition to the LP sequence factor, each routine has its own sequence factor. Both types of sequence factors can be changed dynamically. The relationship between sequence factors and routine execution is shown in Figure 11. In this example, LP0 and LP2 have an LP sequence factor of one. In other words, they get a sync flag on every interrupt by the CP synchronization interrupt. LP1 has an LP sequence factor of two and gets a sync flag on every other loop. Note that routine c is a time-consuming routine and is placed on an LP with an LP sequence factor of two. In the example of Figure 11, routine d may have a routine sequence factor of two.

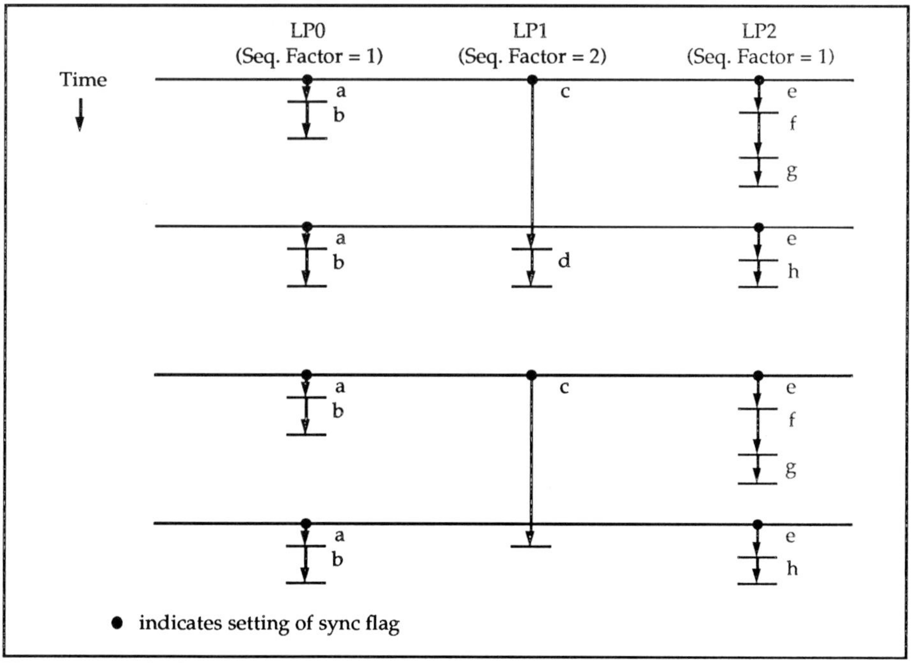

Figure 11.□Application of sequence factor to LP routines.

Because it is assigned to execute on LP1 (with an LP sequence factor of two), this routine executes every four loops.

In Figure 11, LP2 executes code for several routines. Routine f has a routine sequence factor of two and executes on every other time slice. Note that it keeps its order of execution, which is critical in a real-time system. Also note that routines g and h alternately execute. This is an example of a single package that has a time-consuming calculation divided to execute at the loop rate.

Typically, the execution on one LP proceeds without much interaction from the other LPs. However, there may be cases that require a handshake mechanism between routines on different processors. In most of these cases, the handshake is necessary to obtain data at the proper time. In these cases, a short interface routine can execute on one LP to transfer the data to or from other LPs.

4.3 Synchronous common block

The synchronous portion of MICA permits developers tighter control of memory allocation than does the asynchronous portion. Though typically all memory in a tightly coupled system can be accessed by all processors, it is not necessarily true that all of it can be accessed at the same speed. In MICA, partitioning of memory is dictated by the expected

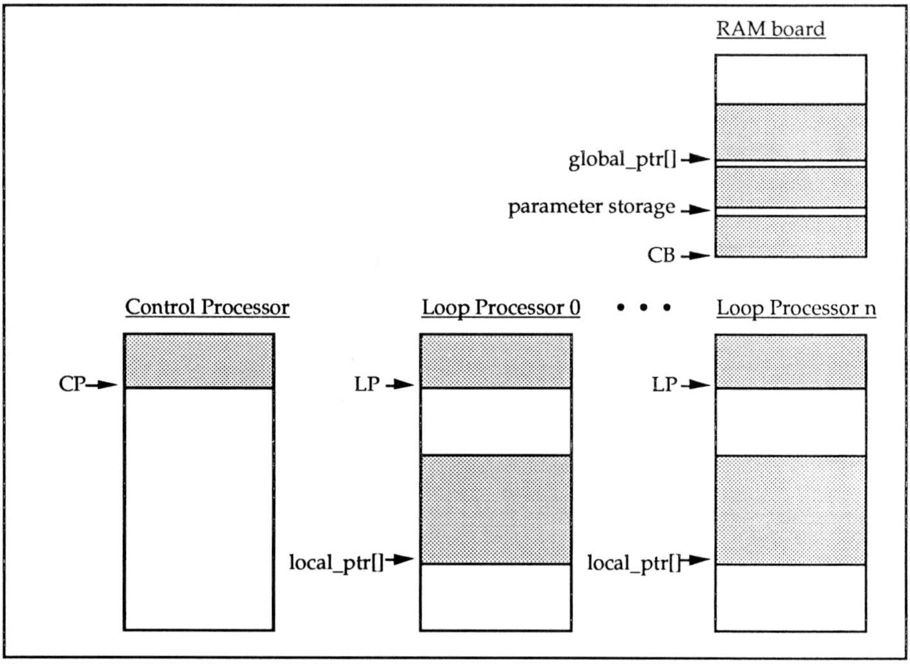

Figure 12. Synchronous common block memory organization is distributed among the processors that handle the cluster processing. Storage is allocated to minimize backplane bus accesses.

number of accesses. Data are assigned to reside on the processor that accesses them the most. A common store is used to contain data that all processors need equally. A simple diagram of a synchronous memory organization is shown in Figure 12. Note that the synchronous common block is separate and independent of the asynchronous blackboard. Asynchronous blackboard accesses (and associated blocking) must not interfere with the operation of the synchronous system. Also note that the memory organization can be customized for the particular objective. For example, parts of the common block may need to reside in dual-ported memory attached to custom control processors.

The CP has one block of memory, pointed to by the CP pointer. The CP structure contains information required by the sync-interrupt routine. This includes the LP sequence factor information used by the interrupt routine. It also contains information on all packages defined for all modules. On each LP, two memory blocks are defined. The LP structures contain the LP sync flags and other LP specific information for each LP. The LP structure also contains the chore vector list and sequence factors and an event queue to allow any LP to post events to asynchronous processes. An array of memory blocks is also assigned by the LPs to contain module-specific local information. These memory blocks are uninitialized data areas that can be used by packages to contain package-specific variables. In an extendible system such as MICA, care must be given to not create too many static global variables that are used by only one package.

The RAM board contains three blocks of memory for use by the MICA synchronous system. First, a common area, pointed to by the CB pointer, is used to contain information about the motors, joints, and other global information. Second, parameters are stored on the RAM board. These parameters are read from the disk system by the CP during initialization and are used when the corresponding package is active. Packages are changed dynamically; therefore, all parameters must be available in memory because there is no time to read parameters from disk when a package is installed. The last memory block to reside on the RAM board is actually an array of memory blocks, one for each module. The global data areas for the modules are similar to the local areas on each LP but are used to communicate package information across LP boundaries and to contain data for other modules.

4.4 Cluster Interface

The synchronous portion of MICA is controlled by an asynchronous interface to allow the asynchronous system to control synchronous loops by using a high-level library interface. For example, high-level events can be passed through the asynchronous, distributed system to command a complex sequence of motions within a cluster. The interface between the synchronous LPs and the asynchronous CP is crucial. The CP must be able to control the operation of any of the LPs but must also be synchronized with the synchronous loop rate. This is handled by an event structure known as "steps." These steps are added to a queue that is read at the start of each LPs synchronous control loop. LPs can also add to the step queue. In this manner, all changes occur at the start of the synchronous control loop in a precise manner.

4.5 Multiple cluster systems

Our initial implementation of MICA contains only one cluster. However, it is desirable to give consideration to expansion above the cluster level in the hierarchy. Figure 13 gives an example of how related clusters may be connected. This is not meant to imply that the clusters are synchronized. An interface package within each cluster would be assigned to control the command and data interface between clusters.

Remotely connected clusters may require synchronization. This suggests using a dedicated communications link. This communication link, which would maintain the synchronization, would have to handle data as well as control information. The LTM[10] is one example of a remotely connected synchronous system.

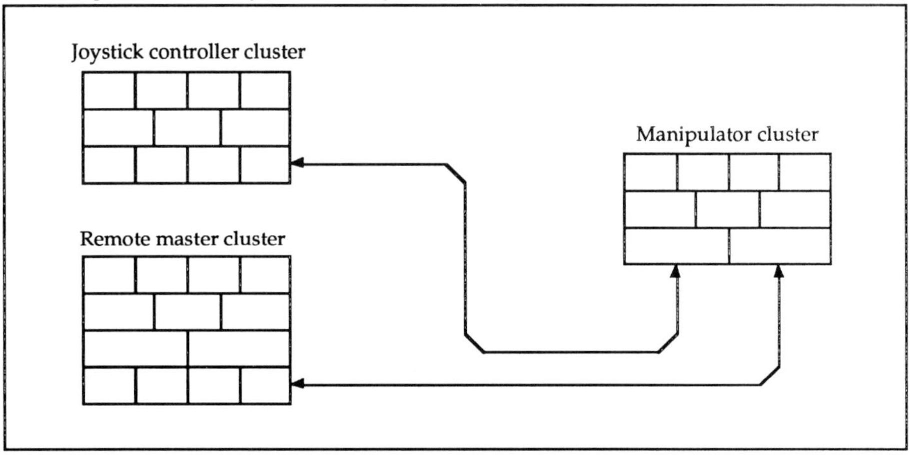

Figure 13. MICA supports complex control systems by enabling multiple clusters to be coordinated.

5. Software Engineering Considerations

A complex control architecture is not very interesting unless it can be implemented. MICA provides not only for capability in the finished system but also for capability in its construction. The differing needs of system designers, developers, integrators, and maintainers are frequently neglected in discussions of system architectures. MICA addresses some (but not all) of these needs by using either simple, but flexible, conceptual models or concrete delivered capability.

5.1 Design

Design is the process that details the method by which a desired complex behavior can be generated. It consists of decomposing the whole behavior into a possibly large number of simpler behaviors. Each of these individual behaviors is simple enough to be realized in the form of a sequential computer program. A designer has much to think about and would prefer to concentrate on the behavioral decomposition rather than implementation details.

MICA assists designers in three ways. First, it flattens out a heterogeneous, distributed memory environment by creating the illusion of a homogeneous system communicating via common memory. Thus, to a first approximation, the designer does not have to worry about the number of processors in the system, their relationship to one another, or the operating system software running on each. Second, by explicitly separating data from control, MICA allows the designer to think in terms of continuous processes that exchange data without synchronization, and ballistic processes that coordinate themselves by exchanging messages. Third, MICA assists in the detailed design by providing a simple, uniform protocol for the exchange of data.

5.2 *Development*

Typically, the development stage in a project involves more people than any other stage. Information that must be shared by all parties in the development is therefore expensive. Furthermore, if the information is difficult to understand, there is a high potential for error.

MICA addresses this situation by providing (1) a small and simple function ensemble that provides all interprocess and intersystem communication services in a uniform format and (2) a simple, uniform, easy-to-use syntax for referencing data. Because the function ensemble is small and simple, error-prone communication between people is minimized. Because the syntax is difficult to get wrong, the potential for implementation errors is minimized. Developers are freed to concentrate on delivering capability in their component rather than being distracted by complex interprocess communication protocols.

5.3 *Integration*

During system integration, it is desirable to be able to test each component of the system independently and to gradually assemble the finished system by assembling and testing increasingly larger subsystems, fixing problems as they are found.

MICA addresses these concerns with the blackboard and event models. The simple, uniform syntax by which references are made provides a simple method for creating scaffolding and monitor processes that can simulate input to modules or subsystems or record their output. Because all I/O in modules not tied directly to hardware is done through the blackboard, MICA makes it possible to move processes around in a heterogeneous system without making changes to the source code.

MICA was designed in particular for robotics applications, which are very often time sensitive. Use of direct memory references to the blackboard, rather than going through intermediate software (as is done, for example, in Carriero and Gelernter[4]), makes these references, particularly reads, relatively inexpensive in terms of time. Perhaps the most objectionable aspect of the system from the performance perspective is the mandatory global broadcast associated with writes. In cases where it is necessary to avoid this delay, it is possible to dynamically disable intersystem communication on a process-by-process basis. This can result in a substantial time savings, and can be exploited with great effectiveness in instances where all consumers of the relevant data are known to be using the same address space as the producer.

5.4 *Maintenance*

MICA was designed in particular to make the software easy to maintain. By "maintenance," we mean not only fixing bugs but also extending and improving the system by adding capability and new modules and porting the system to new environments as the opportunity arises.

The first maintenance issue, finding and fixing bugs, is addressed by the blackboard model. Data written to the blackboard stay there until they are overwritten. The data are available to any process that wants to read them, including in particular diagnostic utilities. A system integrator or maintainer who understands what the correct results should be for a given situation can easily create diagnostic monitors that display what the data actually are. Thus, incorrect results are easy to detect.

Our policy of allowing only one process to write to any given location on the blackboard means that when incorrect data are detected, the offending process is immediately identified. Thus, errors can be isolated rapidly, and we can concentrate effort on a single process consisting of a few hundred to a few thousand lines of code rather than having to search through the entire system for every bug.

Extending the system takes three forms: adding events to an existing process, extending the output of an existing process, and adding new processes. Adding event codes to a process involves no changes outside that process except for a change to the event definitions file. Once changed, the code must be recompiled, as does the code that sends the new event to the process. Enhancing the output of a process or adding a new process involves changes to the blackboard definitions file. MICA imposes an implicit ordering of the Post-Its on the blackboard and an implicit ordering of the data on each Post-It. Therefore, changes to the blackboard definitions file do not interfere with the operation of any existing program as long as new Post-Its are added to the end of the definitions file and changes to the data on a particular Post-It are made at the end of the Post-It. Naturally, once the definitions of the blackboard have changed, any process that wants to communicate by using the new data elements has to be recompiled (but not other processes).

Additions to the blackboard definitions file require that MICA assign new addresses to the Post-Its on the blackboard and generate a new communication system. Because address binding is deferred until runtime, changing the locations of the Post-Its on the blackboard does not require that any currently working code be recompiled or relinked. Because MICA automatically generates all codes that depend on the definitions file, updating the communications system is quick, easy, and correct. This specific aspect of system maintenance provided a large fraction of the original motivation for the development of automatic code-generation capability.

Finally, computers, operating systems, and even robot hardware may evolve at a rate different from that for the software. Therefore, existing code should be easily ported to new configurations as they are brought on-line. MICA addresses these issues by utilizing only the minimal set of hardware and operating system specific services necessary to meet its specification and by isolating these services carefully. Thus, moving MICA from one system to another does not require reinventing the whole system.

6. MICA Implementation

The work described in this paper benefits the U.S. Department of Energy Nuclear Engineering Robotics for Advanced Reactors Program[12]. This program consists of research teams from five institutions. Each institution contributes software and/or hardware to joint technology demonstrations in Oak Ridge, Tennessee. ORNL performs final system integration. This program provides all participants an opportunity to work together in an open environment.

6.1 HERMIES-III system

MICA has been implemented on the HERMIES-III mobile robot[13]. A prominent feature of HERMIES-III is the CESARm manipulator[5]. Also included on HERMIES-III are several sensor subsystems including sonars, a laser range camera, and three charge-coupled-device (CCD) cameras.

Several computer systems control HERMIES-III. One of these systems is a remote Silicon Graphics IRIS, running UNIX, that provides the human-machine interface with the robot. A DEC□MicroVax interfaces to a joystick for teleoperation. These systems are connected with HERMIES-III via Ethernet. An Apple Macintosh is connected to the robot via a serial link. The robot contains seven VME 68020 boards running the Microware OS-9 operating system. Three of these processors run a MICA cluster for control of the CESARm manipulator. Four are available to execute other processes. Also on-board is a 16-processor NCUBE hypercube parallel computer used for image processing and analysis. MICA events propagate to all of these systems. The MICA asynchronous blackboard is not replicated on the NCUBE processors or on the Macintosh.

6.2 CESARm manipulator

The CESARm manipulator has 7 degrees-of-freedom. That is, seven unique actuators that contribute to the motion of the end effector. Since only 6 degrees-of-freedom are required to uniquely specify a point and orientation in space, CESARm is said to have "redundant" degrees of freedom. Complex algorithms[5] are required to resolve the redundancy, smoothly controlling the end effector motion while a cartesian move is being executed. The CESARm manipulator also has coupled actuators. That is, motion in one of the motors affects motion in several of the joints. Another motor controls a complementary motion in the same joints. Therefore, a unique relationship exists between the motor positions and the joint positions. Because of the complexity of the control algorithms, which execute at 150 Hz, two LPs are used to control CESARm.

6.3 CESARm modules

Figure 14 suggests the degree of modularity attained by MICA. This simple diagram, specific to real-time arm control, shows the division between high-level control module, follower module, servo-level module, and input/output module. The output of the high-level control module is directed into the follower module, which performs appropriate transformations on the desired position data. The follower module feeds into the servo-level module, which might consist of a PID controller. However, some connections may not be reasonable, because the 'motor-level' follower is not being connected to the 'torque controller' servo-level package. It is up to the system designers to decide which packages are compatible with each other. Figure 14 also shows that a simulation can be performed simply by choosing the simulation package instead of the actual I/O package. For the CESARm manipulator, several modules have been defined and are described below. Additional modules may be added as needs warrant.

IO_MODULE. The I/O module performs all system input and output. Before any drive output takes place, the system state is checked for errors. If an error occurs, the brakes are locked immediately and action is taken to prevent the error from persisting. This action involves removing the execution vectors for the other modules that may be trying to command an illegal motion.

SERVO_MODULE. The servo module contains the motor-level servo loops for all 7 degrees-of-freedom. Included is a conventional PID controller with some custom digital filtering.

KINEMATIC_MODULE. The forward kinematics (conversion from joint angles into cartesian end effector coordinates) is executed in the kinematic module. Motor, joint, and cartesian velocities are calculated from the change in position.

FOLLOWER_MODULE. The follower module performs desired transformations from one coordinate frame to another. For example, if a cartesian motion is desired, a follower

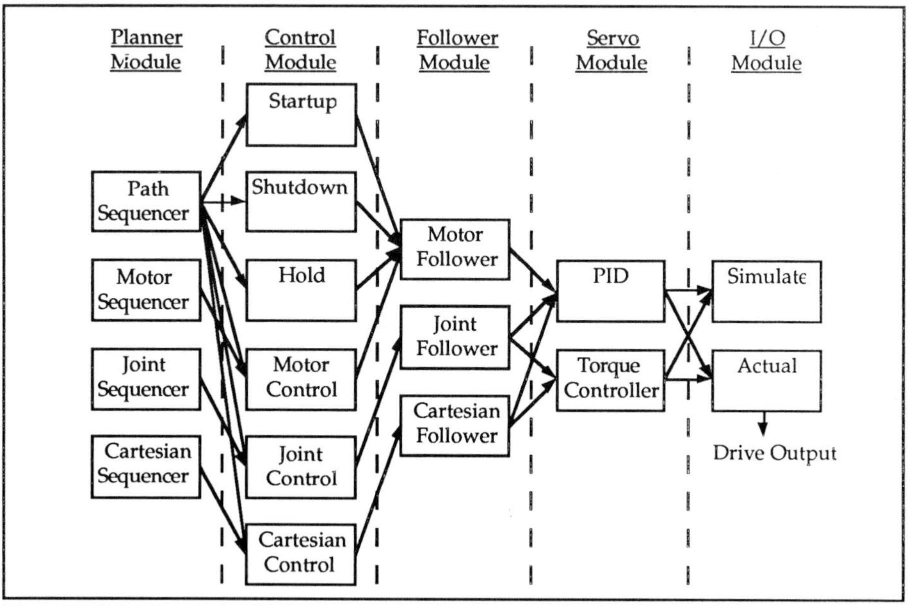

Figure 14. Modular relationship between routines for control of the CESARm. A simulation, for example, can be performed by choosing the Simulate I/O package instead of the Actual I/O.

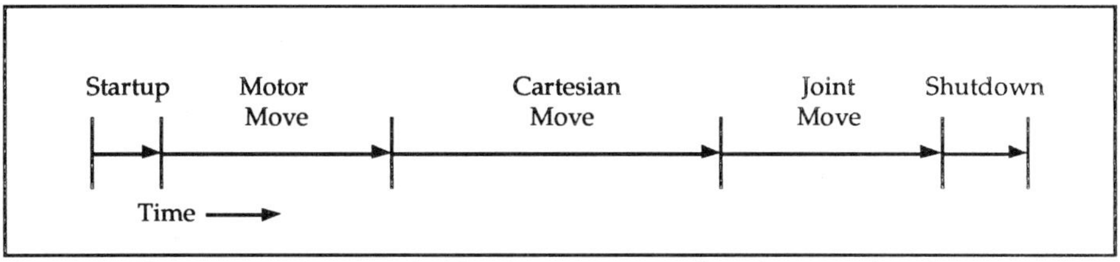

Figure 15.□Planner sequencing example. The planner sequences through any number of control packages, thus allowing complex motions to be performed.

package transforms the desired cartesian input to a motor-level output for the servo module. If a motor-level trajectory is desired, no conversion is required.

CONTROL_MODULE. The control module provides incremental trajectory planning, creating the incremental points that are passed to the follower module. The input to the control module is a distant point that requires some trajectory planning to reach. This planning consists of a speedup period, followed by a constant velocity motion, and ending with a slowdown period so that the ending position is met exactly. One good example of a control package is a cylinder scan package that controls a complex trajectory that keeps the end effector at a constant distance from the center of a cylinder and also keeps the end effector pointing toward the center.

PLANNER_MODULE. The planner module controls the overall trajectory planning, controlling a sequence of primitive moves. These moves are typically input from the asynchronous interface as input from the outside world. However, a complex planner package can determine the sequence of primitive moves at the synchronous level. This determination could take into account the output of collision detection code that changes the desired motion immediately. A requirement is that the planner module be able to stop the primitive moves being executed by the control module. An example of sequencing as controlled by the planner module is shown in Figure 15. In this example, a motor-level move is executed after startup, followed by a cartesian move and a joint move. Finally, a shutdown occurs.

6.4 High-level function library

A high-level function library was created for the asynchronous interface of any process in the system to the CESARm system. An example of a complex motion is shown in Figure 16. The initialize function sends an event to the CP, which clears a queue pointer. All control functions that follow send additional events to the CP, which places a copy of the event in a holding area in the common block. When the finish function is executed, another event is sent to the CP, thus starting the motion.

The arguments to the function calls themselves simply describe the ending position in the desired coordinate frame (motor, joint, cartesian). Additional arguments for the motor and joint moves define the execution time and tolerance of the motion. The cartesian move defines a movement velocity and an optimization criterion for the inverse-kinematic algorithm.

```
#define TOL 100      /* motor position tolerance */
void do_complex()
{
  initialize();
  startup();
  motor_move(0, 28160, 0, 0, 0, 0, 0, 4.0, TOL);
  cartesian_move(-1.2, -.4, 0.5, 0.05, 0);
  motor_move(-14348, 28045, 0, 6635, 172, -180, 87, 5.0, TOL);
  relative_motor_move(1000, -1000, 500, 100, 0, 0, 0, 2.0, TOL);
  delay(1.0);
  motor_move(0, 28074, 3018, 0, 3647, 4130, -2221, 5.0, TOL);
  joint_move(0.0, 180.0, 0.0, 0.0, 0.0, 0.0, 0.0, 4.0, 0.1);
  shutdown();
  finish();
}
```

Figure 16.□Example of complex motion using the function library. These functions build events that go to a planner package, which queues them until told to execute. In this example, the finish function indicated that all sequence events are queued and that motion should begin. When motion is complete, the planner package sends a done event to the process that initiated the motion.

Among the several interesting considerations in this mechanism, one is that the motion functions are nonblocking, allowing an event-driven process to perform other functions. When the motion has completed, an acknowledgment is sent to the process that initiated the motion. Another consideration is that some motions may require a very little time. With these motions queued, no delay occurs in stepping between motion primitives.

6.5 Human-Machine Interface

A human-machine interface (HMI) provides a graphical development interface to the CESARm cluster. Packages can be selected for installation from the HMI and can be used directly to control the CESARm motion. All parameters can be set from various custom interfaces specific to the relevant package. In addition, motor and joint positions can be viewed in a separate status window. Because any event-driven process can command CESARm to perform complex moves using events, the HMI interfaces with the CP by using the asynchronous event system. For CESARm, the HMI is implemented on an Apple Macintosh II computer. Other development platforms could be used, but the availability of the Macintosh made it a logical choice for system development.

A unique aspect of the HMI is that all development for the multiprocessor synchronous system (two LPs) and its asynchronous interface (a CP) can be carried out on the HMI itself without being connected at all to CESARm. The Macintosh used for the HMI deals with its own event loop, the HMI event loop, the CP event loop, and a simulated synchronous LP control loop. This combination creates a stand-alone simulation environment, which is useful in development.

While MICA was being created, this environment eliminated the problems associated with distributed development (Macintosh HMI and the CP/LP VME system). Once the simulated multiprocessor environment was available, CP and LP development progressed rapidly. This environment has proven to be useful, even after the system became operational. Users have been able to develop new packages for CESARm on a stand-alone system, simulating the final results without risking potential damage to the manipulator.

7. Conclusion

MICA establishes synchronous real-time control in tightly coupled multiprocessors and asynchronous real-time control on both local processors and processors connected over a LAN.

The asynchronous portion of MICA can pass control events to any other process in a loosely-coupled distributed system, as well as pass data to other processes through a distributed blackboard scheme. One key feature is that the blackboard scheme placed no constraints on the size and types of data. For example, images can be passed from one process to another.

The asynchronous part is typically applied to real-time situations where strict deterministic processing is not required. Such applications may include passing an image from a robot system with a camera to a remote high-speed computer system with a high-speed processing capability. Results of such an application are typically needed as soon as they are generated, but not required to be synchronized with another process. The synchronous part is applied in applications requiring precise time-synchronized control. Application to the control loops of a robotic manipulator is typical, although other applications are possible as well. A synchronous application, such as a robotic manipulator, can be closely linked with an asynchronous system, such as a tracking system. In such a case, the tracking system could post object location information in the blackboard memory, while the manipulator reads this at a synchronous rate to follow the object.

Another key feature of MICA is it's portability to different types of operating environments. MICA has been applied using the OS-9, VxWorks, Apple Macintosh, and Unix operating systems. This portability between operating systems is a result of the need to use MICA on several projects. The development of MICA has successfully shown that complex robotic systems can be implemented with relative ease. Not only has MICA proven to be successful in the HERMIES III/CESARm implementation, it has also been utilized in an underground storage tank technology demonstration[3] sponsored by the U.S. Department of Energy Office of Technology Development(OTD). In this system, MICA is used to control a SPAR RMS 2500 long-reach manipulator. For this purpose, the MICA control code was ported to the VxWorks operating system. The ease of MICA implementation on another robot system and under different operating system attests to its portability. Approximately 4 man-months of effort were required for the complete transfer and full checkout. During this time, MICA was also enhanced to include new capabilities. Therefore, subsequent applications of MICA should require even less time. The MICA kernel is being evaluated for its potential use as part of a standardized OTD control architecture. All system requirements have been met in the design of both the asynchronous and the synchronous portions of MICA. The MICA modularity has been exploited in the design and integration of both asynchronous and synchronous modules as well as in heterogeneous systems. The extendibility has been useful in adding new capabilities to the HERMIES III robotics system.

A primary benefit of MICA is in system integration and maintenance. Large amounts of complex code have been integrated into a single operational system. Although some integration tasks have not been performed perfectly, the integration philosophy of MICA has proven itself and will be exploited again. Because MICA encourages consistent coding techniques, large teams can effectively work together on final integration. Possible future refinements to MICA include identifying and addressing additional aspects of distributed real-time systems such as remote synchronous communications and control.

8. Acknowledgments

We thank John Jansen, Reid Kress, John Rowe, David Thompson, and Wayne Manges for their helpful comments; Barry Burks, Bill Hamel, Reinhold Mann, and Frank Sweeney for their support and encouragement; and M. Trivedi, M. Abidi, D. Baker, C. Bidlack, M. Beckerman, J. Borenstein, C. Crane, R. Giddings, Y. Koram, W. Macaulay, U. Raschke, D. Reister, G. Reuter, D. Roscheleau, K. Skifstaad, J. Sluder, D. Tesar, S. Tosunoglu, J. Tulenko, D. Wehe, T. Weymouth, and the other participants of the Robotics for Advanced Reactors Program, for their patience, persistence, and comradery. The work described was supported by the U.S. Department of Energy Office of Basic Energy Sciences and Office of Technology Support Programs, Office of Nuclear Energy. This manuscript has been authored by a contractor of the U.S. Government under contract DE-AC05-84OR21400. Accordingly, the U.S. Government retains a nonexclusive, royalty-free license to publish or reproduce the published form of this contribution, or allow others to do so, for U.S. Government purposes.

9. References

1. J.S. Albus, H.G. McCain, R. Lumia, "NASA/NBS Standard Reference Model Telerobot Control System Architecture (NASREM)", NIST Technical Note 1235, July 1987.
2. R.A. Brooks, "A Robust Layered Control System for a Mobile Robot", *IEEE Journal of Robotics and Automation* **RA-2** (1986) 14–23.
3. B.L. Burks, G.A. Armstrong, P.L. Butler, P. Boissiere, "Combined Long-Reach and Dexterous Manipulation for Waste Storage Tank Applications", Proc. American Nuclear Society, San Francisco, CA., Nov. 11–14, 1991.
4. N. Carriero, D. Gelernter, "How to Write Parallel Programs: A Guide to the Perplexed", *ACM Computing Surveys* **21** (1989) 323–357.
5. R.V. Dubey, J.A. Euler, S.M. Babcock, "Real Time Implementation of an Optimization Scheme for Seven-Degree-of-Freedom Redundant Manipulators", *IEEE Transactions on Robotics and Automation* **RA-7** (1991) 579–588.
6. R. Englemore, T. Morgan, *Blackboard Systems*, Addison-Wesley, 1988.
7. S.Y. Harmon, "The ground surveillance robot (GSR): An autonomous vehicle designed to transit unknown terrain", *IEEE Journal of Robotics and Automation* **RA-3** (1987) 266–279.
8. S. Narassimhan, D.M. Siegel, J.M. Hollerbach, "CONDOR: An architecture for controlling the Utah-MIT dexterous hand", *IEEE Transactions on Robotics and Automation* **RA-5** (1989) 616–627.
9. B. Nitzbergg, V. Lo, "Distributed shared memory: A survey of issues and algorithms", *IEEE Computer* **24** (1991) 52–60.
10. J.C. Rowe, P.L. Butler, R.L. Glassell, J.N. Herndon "The NASA Laboratory Telerobotic Manipulator Control System Architecture", Proc. American Nuclear Society Fourth Topical Meeting on Robotics and Remote Systems (1991) 171–178.
11. R. Simmons, E. Krotkov, "An integrated walking system for the ambler planetary rover", *Proc. IEEE International Conference on Robotics and Automation* (1991) 2086–2091.
12. F.J Sweeney, M. Beckerman, P.L. Butler, J.P. Jones, D.B. Reister, "Application of autonomous robotics to surveillance of waste storage containers for radioactive surface contamination", Proc. AI '91: Frontiers in Innovative Computing for the Nuclear Industry, September 15–18, 1991.
13. C.R. Weisbin, B.L. Burks, J.R. Einstein, R.R. Feezell, W.W. Manges, D.H. Thompson, "HERMIES-III: A step toward autonomous mobility, manipulation and perception", Robotica **8** (1990) 7–12.

Integrating Visual Procedures for Mobile Perception

Bruce Draper, Allen R. Hanson, and Edward M. Riseman
Computer Vision Research Laboratory
Dept. of Computer and Information Science
University of Massachusetts
Amherst, MA 01003

Abstract[1]

Experimental real-time systems place significant constraints on the underlying infrastructure and support tools they employ. One of the more important tradeoffs involves efficiency versus flexibility: software tools must be flexible enough to adapt to and support new algorithms and approaches as they are developed, without imposing so much overhead that they produce unacceptable response times. For the University of Massachusetts (UMass) Mobile Perception Laboratory (MPL) -- an outdoor autonomous vehicle -- an attempt has been make to strike a balance between flexibility and efficiency by designing a software environment that supports the integration of visual procedures by providing low-cost mechanisms for data storage/exchange and process control. Once successful approaches have been developed and integrated, the software can be "hardened" into less flexible, more highly optimized code that will not be changed often.

1. Introduction

The convergence of computer vision and robotics has created the need for new kinds of software tools which support the integration of interacting visual procedures and subsystems within the constraints of real-time computing environments. The need is becoming particularly acute as applications in autonomous vehicles and flexible manufacturing become more sophisticated. Although *development* environments for computer vision, such as KBVision™, ImageCalc and Khoros (and soon the ARPA IUE) do exist, these environments are not well-suited to real-time applications. Conversely, real-time environments, such as VxWorks™, do not supply much support for computer vision.

This paper describes the design of an evolving research software environment for the University of Massachusettes' Mobile Perception Laboratory (MPL; see below). The environment is designed to support the integration of disparate visual procedures by providing 1) an efficient database for storing and exchanging visual data and 2) a top-level, behavior-based mechanism for controlling the vehicle. More important than the system itself, however, are the design decisions that underlie it. Designing a software environment for real-time vision involves cost/benefit tradeoffs. On the one hand, because the environment is for research, it should be general enough to support not only currently available visual procedures, but unanticipated future advances as well. On the other hand, one cannot afford a profligate generality: mobile robots and autonomous vehicles must act and react in real time, so their software environments must be highly efficient.

[1]This research has been supported in part by the Advanced Research Projects Agency (via TACOM) contract number DAAE07-91-C-R035, by the National Science Foundation under grant CDA-8922572, and by RADC under contract number F30602-91-C-0037.

The challenge, therefore, is to build an environment that supports real-time vision and robotics without getting "bogged down" with rarely-used features or excessive overhead. This requires making hard choices about what features are necessary to support the intended range of applications. To some extent, the decisions we have made depend on details of the MPL project, but in a larger sense, they are based on fifteen years of experience in building experimental vision systems at the University of Massachusettes [Hanson & Riseman 78; Draper et al. 89], and in building tools to support that research [Kohler & Hanson 82; Brolio et al. 89]. Our current system differs from previous vision environments because the constraints of real-time vision forced us to prioritize needs more strongly than before, and although the details of our implementation will probably change over time, the need to prioritize will remain.

2. The Mobile Perception Laboratory

The Mobile Perception Laboratory (MPL) is a modified U.S. Army HMMWV with computer-controlled steering, acceleration and braking, as shown in Figure 1. MPL has been built as part of ARPA's Unmanned Ground Vehicle (UGV) program to test algorithms for directing cooperating autonomous vehicles under real-world conditions. Because of the complexity and diversity of general autonomous navigation tasks, the algorithms being tested on MPL include many of the basic visual tasks of interest to the image understanding community.

Figure 1. The University of Massachusetts Mobile Perception Laboratory

In particular, MPL was designed to support UMass research in *landmark-based navigation* [Fennema 91, Kumar 92, Beveridge 93], *obstacle detection* and *automatic model acquisition* [Sawhney & Hanson 92; Sawhney et al. 93; Thomas 93; Thomas, et al. 93], and to test vision algorithms for the massively parallel *Image Understand Architecture* (IUA) [Weems, et al., 89]. The landmark navigation task requires that MPL acquire landmarks (immobile objects) from an annotated map, recognize them in an image, and determine the position and orientation of the vehicle relative to the landmarks (i.e. pose determination) in world coordinates. The obstacle detection task, whether on-road or off-road, requires MPL to model the surfaces in front of the vehicle and determine whether those surfaces are flat enough to traverse, and if not to plan and execute a path that avoids them. The model acquisition task is to automatically acquire new landmarks from an initial (and possibly sparse) set of landmarks, to improve the vehicle's map of the environment as it drives through (for use by subsequent vehicles) or, in the extreme case, to build an environmental map from scratch. In addition, MPL also serves as a secondary test vehicle for research being conducted at other sites on road following, off-road navigation and stereo image processing.

Physically, MPL is a modification of the ambulance (996) model of the HMMWV, and is quite similar to Carnegie-Mellon's NAVLAB II. In front it has a glass-faced enclosure for forward-looking sensors, including a color CCD camera for road following and a pair of black-and-white CCDs for stereo. On top it has a pan-and-tilt controllable stabilized platform, capable of rotating through 360 degrees, that carries a color CCD camera with controlable focus and zoom and an infra-red sensor (also with controllable focus and zoom). In the back, MPL contains a complete computer laboratory, including power, air conditioning and facilities for two researchers (in addition to the safety driver). On-board computing resources include a Motorola 68020 processor for directly controlling speed and direction, a Datacube MaxVideo 20 image processor, and a four-processor 340GX Silicon Graphics workstation. Space and power has been provided for a second-generation, 16K processor Image Understanding Architecture (IUA), a massively-parallel heterogenous computer being developed jointly by UMass, Hughes Research Laboratories and Amerinex Artificial Intelligence Corp. [Weems, et al., 89].

3. Software Environment Specifications

The first step in designing a software environment, or indeed any support tool, is to identify the problem to be solved. We did *not* need to build another tool for developing visual algorithms: our researchers were already familiar with KBVision and Khoros, and there was no reason to build another development system. The problem facing MPL is the *integration* of independent real-time visual algorithms. MPL is intended as a testbed, not just of individual vision and robotics algorithms, but of complete perceptual robotic systems. The eventual goal is to build a system that can intelligently execute high-level commands such as "Turn right at the first traffic light, and stop in front of the third house on the left." To execute such a command, MPL will have to combine advanced cognitive processes (such as planning) with high-level visual processing (e.g. object recognition, pose determination) and low-level visual reflexes (e.g. road following, obstacle avoidance).

One way to think about integrating processes as disparate as road following and pose determination is in terms of *behaviors* [Payton, et al. 90; Brooks 87], where a behavior is a pattern of (possibly complex) interactions with the environment. An example behavior is driving, in which a vehicle interacts with its environment by 1) following the road, 2) detecting and avoiding obstacles, and 3) obeying street signs and traffic rules. Each of these three processes is an independent *perception-action* [Arbib 72] process that converts signals from the environment to actions by the robot. Together they form a behavior, in terms of a set of responses to the environment[2]. At one level, therefore, the integration problem is to combine concurrent perception-action processes into coherent behaviors.

Unfortunately, perception-action processes have to be created before they can be integrated into behaviors. Some perception-action processes, such as the ALVINN neural-net-based road

[2]Note that our usage of the term *behavior* is not necessarily compatible with Brooks' [Brooks 87]. He uses the term to refer to simple perception-action processes, whereas we use it to denote a set of concurrent perception-action processes.

follower [Pomerleau 90, 92], are simple enough to be treated as a single software module. More often, however, such processes are too large and complex to be developed by a single researcher/programmer or treated as a unit. In these cases, multiple visual procedures have to be integrated to form a single perception-action process, before that process can be integrated with other concurrent processes to form a behavior.

An illustrative example of such a complex perception-action process is the landmark positioning process on MPL. Landmark positioning gives MPL the ability to estimate its current location and orientation on a map by finding a landmark and determining its pose relative to the vehicle. In order to find a landmark and determine its pose, however, MPL goes through a complex series of actions: it uses color to focus attention on the landmark [Draper, et al., 94], a line extraction algorithm [Kahn, et al., 90] to identify line segments in the selected region of the image, and a hidden-line removal algorithm to identify which parts of the landmark are visible (given its currently estimated position). Finally, it uses a sophisticated matching algorithm [Beveridge 93] to match image lines to model lines, and an iterative refinement technique [Kumar 92] for determining the pose of the landmark from the data-to-model correspondence. Each of these procedures were developed separately by different researchers, with two of them being significant parts of Ph.D. dissertations. In order to be useful, however, all five must be integrated into a single perception-action process.

The challenge, therefore, is to build a software environment that supports the integration of visual procedures (developed in KBVision or Khoros, or acquired from another site) into perception-action processes, and perception-action processes into behaviors. At the same time, this environment must be efficient enough to support real-time applications. Therefore, instead of trying to build tools that address every aspect of integration (with correspondingly high overhead), the two aspects of integration that posed the greatest difficulties were identified: data storage/exchange, and process control. MPL's software environment is therefore built around two simple tools: *ISR3*, a visual database for storing, passing and accessing information, and a *script monitor* for coordinating complex behaviors.

4. ISR3

ISR (an acronym for *intermediate symbolic representation*) has been the name of a series of symbolic databases for vision developed at UMass [Brolio, et. al., 89]. One version, ISR1.5, is now commercially available as part of KBVision (see Section 6.1.1 below), while the most recent version, ISR3, is used on-board the MPL. The ISR databases reflect a belief that computer vision requires more than image-like arrays of numerical data; computer vision depends on symbolic representations of abstract image events such as regions, lines, and surfaces[3] and on mechanisms for efficiently accessing data objects under various types of constraints (such as spatial proximity). Although each version of ISR is a refinement of its predecessor, they all assume that visual procedures operate on symbolic records, called *tokens*, or on groups of tokens, and that visual procedures manipulate tokens both for internal computations and for exchanging data with other procedures.

ISR3 is an in-memory database[4] for C structures. It reads a C header file of structure definitions when it is first initialized, and records the size of the structures (called *tokens*) and the data types

[3] The data exchange format of the IUE reflects a similar belief.

[4] Properly speaking, an in-memory database is called a data store.

and offsets of their fields. ISR3 then establishes a database for the structures in the header file, establishing queues for allocating and freeing tokens, semaphores for preventing simultaneous access (see below), and functions for storing and retrieving tokens. Once one process has initialized ISR3, other processes can attach to it (all tokens are kept in shared memory), making ISR3 a mechanism for data exchange as well as data storage.

ISR is not a generic database, however. Although it supports most generic database operations, it has been optimized for real-time vision research by reducing overhead wherever possible while still supporting those functions most often used in computer vision. In particular, vision applications typically need temporary storage for large numbers of temporary tokens, which they then access by name, feature value or spatial location. Most of these tokens exist for only a short duration, such as the time required to process one image or a short sequence of images, and should then be deallocated, although a few correspond to significant results (critical features, updated maps, etc.) and should persist over time. Consequently, file I/O is not optimized: although data is sometimes saved for later analysis in the lab, most data can be kept in memory and then erased without ever being saved to disk (ignoring the effects of paged virtual memories). Token allocation and deallocation, on the other hand, are critical, which is why ISR3 maintains its own token queues.

As a rule, most visual algorithms operate on sets of tokens rather than on individual tokens. Matching algorithms, for example, compare one set of tokens to another. Therefore most storage and retrieval commands are in the form of set operations, such as a request to access all long, straight lines in the upper corner of an image. Synchronization is provided, not at the level of individual tokens, but of sets of tokens, so that processes may iterate over sets of tokens without having to lock and unlock each token individually. Special facilities for optimizing spatial retrieval over arbitrary sets of tokens are also provided, as are macros for iterating over the tokens in a set, and functions for taking the union, intersection and differences of sets[5].

As an example of how ISR3 has been simplified to reduce overhead while still supporting vision research, consider its synchronization mechanism. One task of a database in a multiprocess environment is to stop processes from accidently overwriting or destroying each other's data. In a typical computer vision application, hundreds or thousands of tokens may be in memory at one time. If multiple processes have uncontrolled access to these tokens and modify them, unpredictable interactions may cause elusive and non-repeatable bugs. On the other hand, it is not uncommon for a process to access hundreds of tokens at a time, for example when a grouping routine looks for pencils of converging lines in vanishing point analysis. If such a process has to lock and unlock a token each time it reads a feature value, protection becomes unacceptably expensive, given the relatively slow speed of semaphores under UNIX, .

A compromise used in ISR3, therefore, was to associate semaphores with sets of tokens. When a set of related tokens is created, for example the set of lines from a single image, a semaphore is allocated to protect those tokens. Any ISR3 function that accesses that set of tokens will first check the semaphore; ISR3 access functions that create subsets of a set of tokens assign the same semaphore to the subset that is used for the parent set. If users access tokens surreptitiously through C pointers[6], they are expected to lock the semaphore before accessing the first token and

---

[5]The complement of a set is not well defined, since there is an infinite universe of possible tokens.

[6]Since ISR3 stores arbitrary C Structures, there is no way to prevent users from accessing tokens without using ISR3 functions, once the user has gotten one pointer into the database.

to unlock it after the last token access. As a result, as long as users do not circumvent its safeguards, ISR3 is able to provide process synchronization with very little overhead.

Similarly, ISR3 provides a level of memory management on top of the UNIX operating system. For non-real-time, file-based systems, memory management is not a critical issue; for continuous, real-time systems, however, it is crucial. ISR3 applications operate in real-time loops, allocating new tokens on each iteration. Memory allocation must be rapid, and must avoid fragmenting memory (as repeated calls to *malloc* would). Memory must be recycled, with space allocated to old tokens being reassigned to new ones once the old data is no longer needed. ISR3 satisfies these requirements by providing token buffers (at a level hidden from the user). Calls to create a token actually allocate one from the buffer for that token type, and freed tokens are returned to the appropriate buffer. For users who store tokens in hierarchies, ISR3 also provides functions for tracing through a hierarchy and freeing all the tokens in it, so that, for example, one can free all the memory associated with an image once the image is no longer current.

5. The Script Monitor

The ISR3 database allows visual procedures to store and exchange data within a single, integrated system. It does not, however, address the problem of process control. Most applications require robots to have several different operational modes or *behaviors*, where a behavior is a set of concurrent processes that determine how the robot responds to its environment. Generally, each behavior is intended to make the vehicle function in some desired way. An autonomous vehicle, for example, may drive down a road until it encounters an obstacle, at which point the road following behavior must be suspended (or suppressed) while the vehicle navigates around the obstacle. Once the vehicle has cleared the obstacle, road following may continue. Interrupting one set of processes and shifting to another in response to events in the environment is the task of the script monitor.

Instructions such as "take the first right and stop in front of the third house on the left" must be translated into a *script*, i.e. a partially developed plan, either by an AI-style planner or a human [Fennema 90]. The script decomposes the instructions into a sequence of executable behaviors, where a behavior is defined as a concurrent set of perception-action processes. The task of the script monitor is to execute this pre-planned script, starting up and killing perception-action processes as needed. For example, given the instructions above, the first behavior would be to drive while avoiding obstacles and looking for traffic lights, a combination of three perception-action processes. The second behavior would be to turn right, while the third behavior would be driving while recognizing and counting houses. Once a behavior has been started by the script monitor, it countinues until one of its perception-action processes raises a condition, such as when the traffic light is spotted or the right turn has been completed.

When the environment is well-behaved, behaviors continue until their pre-planned termination condition is reached (such as finding the traffic light); the script monitor's task is then simply to proceed from one step in the plan to the next, killing the processes of the completed behavior and starting up processes for the new behavior[7]. Unfortunately, as every roboticist knows,

[7]The script monitor invokes and kills processes, but leaves the actual scheduling to UNIX. A better approach would be to implement a real-time scheduler at the level of the operating system, but this is beyond the scope of the MPL project. Fortunately, a related lab at UMass is developing the Spring Kerrnel, a real-time operating system that when completed will serve this purpose [Stankovic, et al. ?]

environments are rarely well-behaved. Perception-action processes may encounter a variety of situations which cause them to fail. For example, while driving to the traffic light, the obstacle detection process might find an obstacle blocking the road, or the road following process might become unable to follow the road any further, perhaps because of an intersection. In either case, the behavior is terminated by a condition other than the one predicted in the simple, linear plan, which is why scripts are more than just simple plans; they indicate, for every condition that a perception-action process can raise, what behavior the robot should respond with.

In essence, scripts are finite state machines (FSMs), in which the states are behaviors (i.e. sets of perception-action processes) and the transitions are process conditions. If the robot encounters an obstacle while driving, for example, it enters into an avoidance behavior until it is past the obstacle. Then it returns to its original driving state until another condition (such as finding the traffic light) occurs.

There are, of course, more sophisticated models of computation than finite state machines that could be used to direct a robot's behaviors. FSMs, however, can be easily implemented as look-up tables, creating highly efficient control systems with very low overhead. They can also be analyzed as Discrete Event Dynamic Systems (DEDS) [Ramadge 89] when formal descriptions of the component processes are available. Furthermore, because they are simple, we hope to build automatic planners that generate scripts in response to natural language instructions, creating a fully automatic system with no human intervention.

Perhaps the best reason for implementing control in terms of FSMs, however, is that we have not yet encountered any reason to believe that a more complex mechanism is necessary. Although we do not yet have a natural language interface, we are currently experimenting with scripts built out of six component perception-action processes: road following (ALVINN [Pomerleau 90, 92], servo-based path following [Fennema 90], stereo-based obstacle detection, reflexive obstacle avoidance, harmonic-function path planning [Connolly 92] and landmark-based positioning [Draper, et al., 93; Beveridge 93; Kumar 92]. These scripts control the MPL in real time as it drives around the UMass campus. Although our experiments are still in the early stages, we have so far found FSM-like scripts to be quite effective.

6. Vision Procedure Development

Because a number of software systems are available to aid researchers in developing new visual procedures, work in support of the MPL project has concentrated on integrating existing visual procedures. Nonetheless, the first step in building an integrated system is to develop the component procedures, and in this section we briefly discuss the strengths and weaknesses of the two development environments used in the MPL project, KBVision and Khoros.

6.1 KB Vision

KBVision [Williams 90] is a computer vision software development environment built by Amerinex Artificial Intelligence Corp.(AAI), and motivated to a significant extent by the knowledge-based research at UMass. Although KBVision supports the development of visual procedures in several ways, its most unique and salient features are the ISR visual database (version 1.5) and the Image Examiner graphics tool.

KBVision's ISR is different from the ISR3 described above in several ways: it does not store native C structures, its grouping mechanisms are less general, and it does not provide interprocess synchronization. The most significant difference, however, is that while ISR3 is (properly speaking) an in-memory data store, ISR1.5 is a true database system, using files to transfer data from one procedure to the next. Although slightly less efficient, the database model is far more appropriate for algorithm development than an in-memory store. It allows visual procedures to be developed in relative isolation from each other, and to be tested on data collected from other locations. (For integration, of course, the more efficient in-memory model of ISR3 is superior.) Many of the procedures on MPL were originally developed in KBVision using ISR1.5, and then "upgraded" by replacing ISR1.5 commands with ISR3 commands.[8] KBVision's ISR1.5 can therefore be viewed as the development analogue of MPL's run-time ISR3, and porting visual procedures from KBVision to MPL is a fairly straightforward task.

The Image Examiner provides vision researchers with graphical support for displaying images and data. As such, it is only one of many tools available for examining images. What distinguishes the Image Examiner from most other tools, however, is that it is integrated with ISR and includes routines for displaying most common forms of symbolic tokens. It is therefore trivial, for example, to overlay a set of lines on top of an image, or on top of a symbolic image region. Tokens can also be pseudo-colored according to feature ranges, allowing the user to visually inspect symbolic data, or selected by a mouse. This makes the Image Examiner a much more useful tool than graphics systems that have not been integrated with a visual database.

6.2 Khoros

Khoros [Argiro 91], although often compared to KBVision, is in fact a very different type of system, designed for different uses. Like KBVision, Khoros is organized around a library of visual procedures. But whereas KBVision was designed as a high-level image understanding environment, with facilities for reasoning about lines, surfaces and other symbolic events, Khoros was designed as an image processing environment. As a result, it has no equivalent to the ISR and manipulates only image data. (Khoros provides an *editimage* that is similar to the KBVision Image Examiner, but it applies only to image data.)

What makes Khoros effective, however, is its user interface[9]. Khoros provides a graphical execution environment, called *Cantata*, in which visual procedures are represented by graphical icons. Clicking on an icon pops up a menu of the procedure's parameters, and allows the user to adjust those parameters without having to recompile the procedure or the system. Visual procedures are sequenced by graphically connecting the output of one procedure to the input of the next; loops and branching sequences of procedures can also be graphically created. As a result, it is relatively easy to combine seperate procedures into perception-action processes in Khoros, if one can get symbolic data from one procedure to the next.

7. Conclusion

---

[8]Unfortunately, ISR1.5 and ISR3 are not upwardly compatible, so software modules must be converted from one to the other by hand.

[9]According to AAI, a new graphical user interface for KBVision will soon be released.

MPL's software environment is designed to support the integration of real-time visual procedures with as little overhead as possible. Based on past experience, it was felt that the two critical areas that needed support were data storage/exchange and process control. The heart of the system, therefore, is the ISR3 database, which provides a central repository for all visual procedures running on MPL, and therefore a method by which they can share data. Perception-action processes that convert sensory data into physical or cognitive actions are built by sequencing visual procedures together, and behaviors are collections of concurrent perception-action processes. MPL's script monitor is a low-overhead control system that switches from one behavior to the next in response to conditions, as specified by a script.

8. Acknowledgements

The system and visual procedures described in this paper were developed by the staff and students of the UMass Computer Vision Lab. Although many people contributed to this effort, we would like to give special thanks to Benny Rochwerger and Claude Fennema for their work on the script monitor, Gokhan Kutlu for his work onISR3, Jonathan Lim for re-implementing the geometric model matching procedure, Katja Daumueller for improving the fast line extraction procedure, Shashi Buluswar for implementing the color-based focus of attention procedure, Alan Boulanger for implementing the hidden line removal procedure, Sumit Badal for designing and implementing the stereo-based passive obstacle detection procedure, Srinivas Ravela for designing and implementing the reflexive obstacle avoidance procedure, and Brian Pinette for porting the harmonic planning procedure and helping to calibrate the system. We would also like to thank Dean Pomerleau for providing both his ALVINN code and his assistance, and Carla Brodley for her advice on the focus of attention algorithm.

9. References

Argiro, J., KHOROS Documentation, University of New Mexico, 1991.

Arbib, M. *The Metaphorical Brain: An Introduction to Cybernetics as Artificial Intelligence and Brain Theory*, N.Y.: Wiley Interscience, 1972.

Beveridge, R. *Local Search Algorithms for Geometric Object Recognition: Optimal Correspondence and Pose*, Ph.D. Thesis, University of Massachusetts (Amherst), 1993. Available from the Department of Computer Science as technical report 93-71

Brolio, J., B. Draper, R. Beveridge, and A. Hanson. "The ISR: an intermediate-level database for computer vision.", *Computer*, 22(12):22-30 (1989).

Brooks, R. "Intelligence Without Representation," *Proc. of the Workshop on the Foundations of Artificial Intelligence*, Cambridge, MA: MIT Press, 1987.

Connolly, C. "Applications of Harmonic Functions to Robotics," *International Symposium on Intelligent Control,* Glasgow, Scotland. August, 1992.

Draper, B., J. Brolio, R. Collins, A. Hanson and E. Riseman. (1989). "The Schema System," *IJCV*, 2(3):209-250.

Draper, B., S. Buluswar, A. Hanson and E. Riseman. "Information Acquisition and Fusion in the Mobile Perception Laboratory," *Proc. of Sensor Fusion VI*, Boston, MA., Sept. 1993. pp. 175-187.

Draper, B., C. Brodley and P. Utgoff. "Goal-Directed Classification using Linear Machine Decision Trees," *PAMI*, to appear (1994).

Fennema, C., *Interweaving Reason, Action, and Perception*, Ph.D. Thesis, University of Massachusetts, 1991; available from the Department of Computer Science as Technical Report 91-56.

Hanson, A., and E. Riseman. "VISIONS: A Computer System for Interpreting Scenes", in *Computer Vision Systems*, Hanson and Riseman (eds), N.Y.: Academic Press, 1978. pp. 303-333.

Kahn, P., L. Kitchen, and E. Riseman. "Real-Time Feature Extraction: A Fast Line Finder for Vision-Guided Robot Navigation," *PAMI* 12(11):1098-1102 (1990).

Kohler, R. R. and A.R. Hanson. "The VISIONS Image Operating Systgem", Proc. 6th ICPR, Munich, Germany, October 1982.

Kumar, R. Model Dependent *Inference of 3D Information from a Sequence of 2D Images*, Ph. D. Thesis and Technical Report TR92-04, Department of Computer Science, University of Massachusetts (Amherst), 1992.

Payton, D., Rosenblatt, K., and Keirsey, D. "Plan Guided Reaction," *SMC* 20(6):1370-1382.

Pomerleau, D.A., Gowdy, J. and Thorpe, C.E. "Combining Artificial Neural Networks and Symbolic Processing for Autonomous Robot Guidance," Proc. of the DARPA Image Understanding Workshop, San Diego, CA, Jan. 1992. pp. 961--967.

Pomerleau, D. A., Neural Network Based Autonomous Navigation, in Charles E. Thorpe, Ed., *Vision and Navigation: The Carnegie Mellon Navlab*, Chapter 5. Kluwer Academic Publishers, 1990.

Ramadge, P. and W. Wonham. "The Control of Discrete Event Systems," Proc. IEEE 77:81-98 (1989).

H. Sawhney, R. Kumar, A. Hanson, E. Riseman, Landmark-Based Navigation - Model Extension and Refinement, Proc. DARPA IUW, Washington, DC, April 19-21, 1993.

Sawhney, H. and A. R. Hanson, Tracking, Detection, and 3D Representation of Potential Obstacles using Affine Constraints, Proc. of DARPA IUW, San Diego, CA, 1992, pp. 1009-1017.

Thomas, J., A. Hanson, and J. Oliensis, Understanding Noise : The Critical Role of Motion Error in Scene Reconstruction, Proc. ICCV, Berlin, Germany, May 11-13, 1993a, to appear.

Thomas, J., "Obtaining the Robot Path Using Automatically Acquired Models, Proc. International Conference on Intelligent Autonomous Systems (IAS), Pittsburgh, PA, February 1993b, to appear.

Weems, C.C., Levitan, S.P., Hanson, A.R., Riseman, E.M., Shu, D.B., Nash, J.G., The Image Understanding Architecture, *IJCV*, 2, 251-282 (1989).

Williams, T. Image Understanding Tools, Proc. 10th ICPR, Atlantic City, NJ, June 1990, pp. 606-610.

THE VISION PROGRAMMERS WORKBENCH (VIPWOB)*

NIELS O. S. KIRKEBY and HENRIK I. CHRISTENSEN
Laboratory of Image Analysis, Alborg University
Fr. Bajers Vej 7, Bldg. D1, DK-9220 Aalborg East
Denmark

ABSTRACT

VIPWOB is a workbench intended for modular processing i.e., it launches and takes care of the communication between programs which are part of a system. VIPWOB is primarily for systems of programs or when distributed data processing is required, i.e., using dedicated hardware. It is intended for image processing but since there are no dependencies on data formats, it may be used whenever a skeleton system is required.

1 Introduction

The development of systems for experimental computer vision may be divided into two different approaches: a modular approach and a dedicated approach. The dedicated approach is basically system development from scratch while the modular approach assumes the existence of some sort of skeleton upon which each module in the system is based. To point out the major advantages of each approach they are:

- Dedicated systems usually has a relative long development time but the payoff is in the performance of the final system.
- Modular systems has a relative short development time since they to a large extent are based on existing modules. However, they usually has a poorer performance than their dedicated counterparts since generality often introduces a performance overhead.

Sometimes a combined approach is used: a prototype system is designed, developed and debugged using the modular approach. Later the modular system is rewritten

*This research project has been funded by the Danish Technical Research Council through the "MOBS-rammeprogram" and the EEC through the Basic Research Action BR-3038/7106-VAP

into a dedicated system in order to improve performance. This way one might get the best of both worlds.

Experimental science is an important part of basic research since:

- new theories should prove them self through experiments. Otherwise they are "just" theories.
- expriments might lead to formulation of new theories; i.e., the result of experiments might be generalized.

Thus there is a connection between experiments and theories or knowledge in science.

Vision is no exception to this rule. Experimental computer vision plays an important role in vision. However, experimental computer vision often involves usage of well known algorithms, repetitive usage of software and/or dedicated hardware. Thus experimental computer vision is an obvious area for modular programming, especially since performance often is secondary.

The concept of reusing modules or modular programming is by no means new in experimental computer vision. However, some systems lacks the facility of distributed processing [1] while others imposes a strict object communication scheme. [2] As might be recognized the overall philosophy of the work presented in this paper is to form a very general but simple to use, basis for modular programming.

1.1 Requirements

In order to provide a general basis for modular programming, any type of system architecture should be possible. There are basically three different types of system architectures in experimental computer vision: pipeline, blackboard and directed graph architectures. These architectures are shown in figure 1. The former

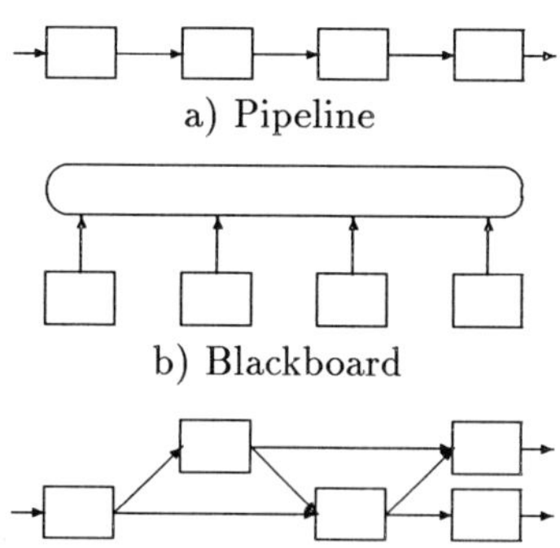

Figure 1: Basic architectures.

two might, however, be implemented as special configurations of the later. The pipeline architecture is probably the most well known. It corresponds to a sequential application of operators. In its most extreme form the blackboard architecture is the complementary to the pipeline. In this case there is a set of modules or operators analysing the same set of data. The directed graph architecture might be viewed as a generalization or mixture of the pipeline and blackboard architectures. In the architecture nodes represent modules and arcs represent communication links. Thus an obvious choice of an architecture for a general basis should be the directed graph architecture. In practice the directed graph architecture implies that each module in the system should be able to handle multiple inputs and outputs. Modules should be also be able to handle multiplexed data; i.e., an output should be able to replicate its data to several inputs and an input should be able to pick up data from several outputs.

Some applications might require special hardware or a distribution of load onto multiple CPUs might be desirable. Thus, a general basis should support execution of modules on multiple hosts. This means that the hosts should be interconnected by some sort of communication medium.

Communication between modules may be divided into the following categories:

- Point to point data flow
- Client / server communication
- Signaling or explicit synchronisation
- Sharing of data structures (blackboard systems)

Point to point communication is the most frequently used type of communication. It provides a memoryless channel for sequential data transfer from one module to another.

The client/server communication is another commonly used scheme. A module (the client) requests data from another module (the server) and once data is available a reply with the data is returned to the client. The communication is thus bidirectional and asynchronous. Asynchronous means that several requests might be sent to the server before replies are returned and the order of replies may be different from the order of requests. The client/server communication model may for example be used for inqueries about module parameters, setting of module parameters etc. Usually the requests are rather limited in size while the replies may be of arbitrary size.

The signal or explicit synchronisation is included here as it is often needed in systems which operates on sequences of data. Explicit synchronisation may for example be used to signal that a new image has been acquired, that some processing has been completed or synchronisation between two parallel data-streams ex., synchronisation between the left and right modules in a stereo reconstruction task. In principle a signal might be viewed as an one-element message but in contrast to messages which usually are queued, the signal is attended to as soon as it arrives and thus it has a higher priority than messages and point to point communication. A special problem in distributed processing is signalling across different processors.

The final category of communication methods is shared data. Sometimes it is inconvenient to use a sequential communication stream, as the receiver might need to access data in an order different from the order in which data are sent. In such situations it might be convenient to have a common data area which allow temporary storage of data and random access to data. Traditionally this is achieved through use of shared memory. This, however, assumes that all processes have access to the same physical memory, which rarely is the case in a distributed system. Consequently facilities should be provided for distributed shared memory.

1.2 VIPWOB Characteristics

In 1988 a consortium consisting of three danish laboraties (Aalborg University, Copenhagen University and The Danish Technical University) decided to initiate a joint effort to specify, design and implement an architecture that would allow sets of modules to be interconnected to form vision and image processing systems.

The objectives for this effort were:

- Design of a testbed or environment that would allow simple construction of application systems.
- The design should allow reuse of as many modules as possible.
- The architecture should allow use of existing software to the extent possible.
- Distributed processing should be possible.
- Easy reconfiguration

From these objectives a software architecture has been designed and implemented. The system has evolved from an initial system which only implemented the most basic functionality of communication to the present version, described in this paper.

In the initial design of VIPWOB the focus was on the interconnection of modules, both new modules and existing modules. Thus the use of a standard data representation was not possible. Consequently the directed graph architecture was adopted as the type of architecture which should be implemented in VIPWOB. This choice was made as it imposes a minimum of structure on application programs i.e., it allows implementation of architetures like pipeline and blackboard systems (using a common database server).

However, there is a conflict between the lack of standard data representation and the requirement that the directed graph architecture should allow multiplexed data. Since no data representation standard has been imposed on VIPWOB, each communication link is viewed as a sequence of bytes. Thus VIPWOB has no sense of the size of the objects being transferred through a communication link. This means that object might get intermixed if multiplexed data were allowed. However, the problem of multiplexed data might be solved by using the client/server model instead.

It was decided that multiple hosts should be connected by an Ethernet network in order to facilitate distributed processing. However, facilities has been provided in the present version to allow implementation of other communication media such as shared memory.

Modules may communicate using point to point communication, messages (the implementation of the client/server model) and signaling. The use of shared data structures has not been designed into VIPWOB but with the directed graph architecture this should not be a problem (Not implementing it in VIPWOB but designing a shared memory system with VIPWOB).

The initial design of VIPWOB allowed use of existing modules without recompilation, that is, only the binaries should be available. The present version does, however, require recompilation and access to at least part of the source code. This is described in more detail later.

Configuration of VIPWOB is done by specifying modules and point to point communication in a text file, which is straight forward to write. There are no graphic design interface as implemented in the majority of other systems as this is viewed as a luxury. On the other hand the system is dynamically reconfigurable. That is modules may launch other modules.

Finally it was decided that VIPWOB should be written in C and be available on UNIX as this probably is the most common system on most scientific research laboratories.

1.3 Summary

To summarize the characteristics of VIPWOB:

- Written in C and available on UNIX.
- The Ethernet network is supported as the communication medium facilitating distributed processing.
- No specific data format is imposed on data communication.
- The directed graph architecture is adopted.
- Relative small changes to existing source code.

2 Overview of VIPWOB

2.1 System architecture

The kernel in VIPWOB consists of three special modules, named the 'shell', the 'master server' and the 'slave server'. The shell and the master server only exists in one instance while the slave server may exists in several instances. Finally there are application modules which are modules written by an application programmer. For the remaining part of this paper application modules will be referred to as modules.

The shell is basically the interface between the a user and the system. It accepts a configuration description and translates this description into a set of commands to be sent to the servers. It is also possible for modules to use the shell to communicate with the user; i.e., modules may send requests to the shell about getting input from the user at runtime. Finally the user may use the shell to send messages to servers and modules. This way the shell also acts as a simple debugging tool.

The master server has two purposes: it contains the global storage of VIPWOB and it can launch slave servers. The global storage is a set of variables which may or may not have values. Both variables and their content are stored as strings exactly like environment variables in Unix. Some variables are special, e.g. there is a unique identifier variable. Thus if a module needs a name for a variable it may access this variable to get a unique name.

There is a variety of ways to access the global storage, some of which will be described in more detail later.

The slave server is the local controller of modules in the system. Basically it can launch modules and send signals to modules.

The shell is the initial program executed by the user. It in turn launches the master server as a subprocess. The master server then launches any slave server needed on local or remote hosts. Finally modules are launched by the slave servers.

The internal architecture of VIPWOB is a tree-like structure with three special purpose modules like in figure 2. Note that this is the internal structure of VIPWOB.

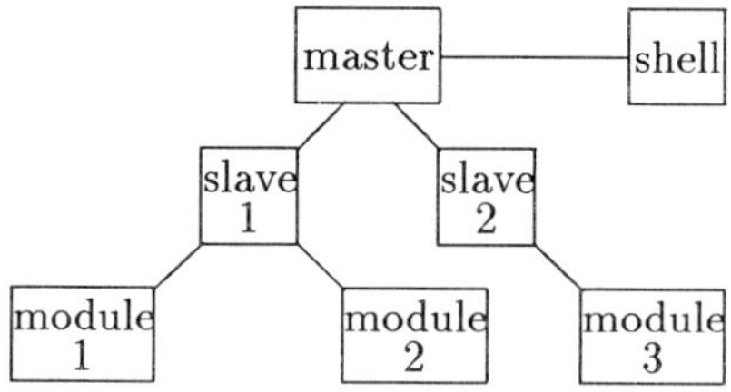

Figure 2: Example of VIPWOB system structure

It is *not* the system architecture.

2.2 *Communication*

In VIPWOB there are three different types of basic communication methods:

- Point to point channels
- Messages
- Signals

Point to point channels are explicit data channels between two modules. Such channels are dynamically configurable, that is they may be created or destroyed during runtime. They are implemented by overloading the functions *open*, *close*, *read* and *write* of the standard Unix C-library. Because of this, they appear as normal Unix filedescriptors to the application programmer. Point to point channel are only used by modules, no server uses them.

It is the point to point channels which "defines" the system architecture. However, since these channels are dynamically configurable, the system architecture is in fact dynamically configurable. Point to point channels may not even be defined at system startup but added later. This way a module may in fact decide what kind of architecture the system should use.

The messages of VIPWOB implement the client/server model. A message consists of a header and a (possible zero length) data block. The header of a message contains information about the destination of the message, the source, priority, whether the message is a request or a reply, the length of the data block etc. Unlike point to point communication a special set of functions must be used in order to use messages.

Since all parts of VIPWOB communicates by means of messages, like an ordinary network of computers, each part is assigned an address. An address consists of two numbers: a major address and a minor address. The master server and the shell has fixed addresses, 0.0 and 0.1, respectively. A slave server has always a minor address equal to 0 and a module has always the same major address as the slave server controlling it.

All communication with the shell, the master server and the slave servers are command strings send as messages. There are no binary data involved. Hence these commands are independent of CPU-dependent parameters like endian-type, the size of an integer, etc. However, messages between modules don't have to be strings, they may contain binary data. VIPWOB does not impose any format of the data; it just copies the data.

Finally there are signals. Since VIPWOB is a distributed system, signals are implemented as a command to a slave server. E.g. if module A want's to send a signal to module B, module A must send a "signal" command to the slave server controlling module B. Signals are only used by modules, no server uses them, except that they originate from a slave server.

3 Module Programmers Interface

To use VIPWOB the modules must use a special VIPWOB library. This library contains code to communicate with the servers. The usage of this library does, however, not imply that programs linked with this library cannot run as "standalone" programs. It is possible to execute such programs from the standard Unix shell when the programmer obeys certain restrictions in the usage of the library.

Basically this library provides a general interface around the module. Thus a module looks as in figure 3. This approach has the advantage that all interfacing like

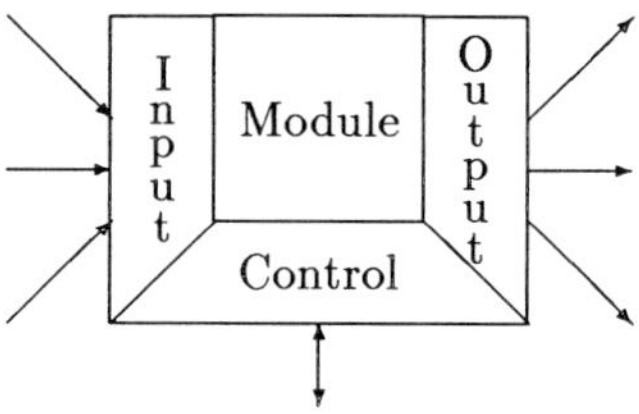

Figure 3: Module interface.

initialization, the implementation of communication, handling signals etc. are hidden from the application programmer.

3.1 Initialization

For the modules to work with this library, a call to a function named **VWInit()** **must** be made in the **main()** [1] function of these programs and it must be one of the first functions called, especially before *any* I/O functions are called.

Thus the scheme in the **main()** function looks something like this:

```
main(...) {
  /* Defining variables */

  VWInit();

  .
  /* rest of code */

  .
}
```

Calling **VWInit()** does not imply that programs have to run under VIPWOB. They may be used as regular programs. However, this is only the case if **VWInit()** is the only VIPWOB function called. The return value of **VWInit()** tells whether the program is launched by VIPWOB or the standard shell. This way a program can

[1] main() is the normal entry function in an C-program.

decide to run in two different modes: when launched inside VIPWOB it can use the features of VIPWOB like messages etc. and when launched from the standard shell it refrains from using them and uses other methods instead (or works in an entirely different way). Note that since point to point channels in VIPWOB are implemented by overloading the functions *open*, *read*, *write* and *close* they appear just like any ordinary *filedescriptor* in Unix. Thus programs which use either low level I/O (in Unix terminology) or buffered I/O on the standard filedescriptors only needs to call **VWInit**() in order to be used by VIPWOB.

3.2 Communication

Point to point channels

As already mentioned point to point channels are implemented by overloading the basic I/O functions of Unix. Thus they are described by an *filedescriptor*. This implies that to send or receive data on point to point channels is just like writting to or reading from a file. Ex. to read data from a channel the function *read* is used:

read(fd,buffer,size)

where *fd* is the filedescriptor. More about this later.

Point to point channels may be opened during runtime by using various function calls or they may be specified in a configuration description in which case they exists when **VWInit**() returns just like the standard filedescriptors of Unix (standard input, standard output and standard error output) exists when a program is executed.

Messages

Messages are *the* fundamental communication method of VIPWOB. Point to point channels are mainly a concept in the module library. Messages are used to communicate with servers and other modules.

Messages are basically divided into two groups: requests and replies. Request messages are assigned an unique identification number (id) when they are "created". Replies are always related to a specific request and has the same id as the original request. Thus the idea is to send a request message and get hold of it's id, then wait for a reply which has the same id. In this scheme replies are always returned to the source of the request.

Modules has a further division of messages. They are divided into system messages and module messages. System messages are intended for the interface part of the module, that is the module library, while module messages are forwarded to the application.

Messages are queued when they arrive to a module or server. The order in which they are queued is usually the same as the order in which they arrive. There are, however, some mechanisms which may change this order such as priority. One should also note that replies and requests are put in the same queue. Thus it is possible to retrieve a reply in two ways:

1. By waiting for a specific reply to arrive and extract this reply from the queue. That is the reply don't have to be at the end of the queue to be retrieved; they may be anywhere in the queue.
2. By "popping" off messages from the queue. This way a mixture of requests and replies are retrieved from the queue.

Note that requests can only be retrieved in the order they are put onto the queue.

Messages has attributes assigned to them. These attributes are used to describe the message whenever it enters a module or a server. The attributes are:

system This attributes is used to distinguish between messages meant for the module core and messages meant for the module control. This attribute has no meaning to servers.

reply This indicates that the message is a reply to a previous request.

acknowledge This indicates that the receiver of the message should return a reply in all circumstances even when there is no data to return. This is a way to solve the problem of waiting for success or failure. Ex. when a module receives a request to perform some action there might not be anything to return in the case of success but if an error occurred a reply is sent. Now, the sender of the request should eventually wait to see if an error occurred but in the case of success no reply is returned. This attribute is used to tell that the sender wants a reply whether the action was successfull or not. Note that it is the module programmers responsibility to return a reply. Also note that this attribute only has meaning to request messages, not replies.

All servers obey this attribute.

urgent This attribute indicates that this is a high priority message. That is the message (whether it is a request or a reply) is put in front of the input queue. Of course if there is another message with this attribute in the queue, the new message is put into the queue after this message. This might be viewed as if there are two input queues: a queue for messages with the urgent attribute and a queue for messages without it. When the module wants to retrieve a message the queue with "urgent" messages are checked first and only if it is empty, the second queue is checked.

Note that this attribute applies to both requests and replies. That is the reply of an "urgent" request message has also the urgent attribute set.

warning This attribute is mainly for replies to indicate that it is a message containing a warning of some sort.

error Like the warning attribute but used to indicate errors.

shell This attribute only has meaning to the shell. Normally the shell just displays all messages that arrives to it. This attribute is used to indicate that the shell should accept this message as a command.

debug This attribute is for debugging purpose only. When it is set each server or module the message parses through when it is routed displays some information about the message according to what the server or module is told to display.

multi Since there is a limit to the size of the data block of a message (currently 64 Kbytes) this attribute has been introduced. It is used to indicate that this message is part of a set of messages. All messages in this set has the same id but different sequence number. Note that the entire control mechanism to handle messages with this attribute is the responsibility of the module programmer. The only thing VIPWOB is capable of is **not** to update the message id.

Above it was stated that messages are queued when they arrive to a module. There are, however, a possibility to setup a message interceptor, that is a function to be called each time a message arrives. This function may choose to react on the message immediately and thus the message is not queue or the function may choose to "ignore" the message and then the message is queued.

Even though messages only contain a single destination specifier it is possible to "broadcast" messages, either to a selected group of receivers or to all modules. Groups are maintained by the master server and created, modified or deleted through a set of commands from the modules. Thus it is the modules responsibility to create a group. A group is a specific address which is translated into a set of addresses. Every address of this set receives a copy of the message. Another method for broadcasting is by specifying "wildcard" addresses. I.e. when sending a message to the address 1.-1 every module with the major address 1 receives a copy of the message.

Signals

VIPWOB uses the signal environment of Unix to implement signals. That is signals are in fact ordinary Unix signals. However since VIPWOB is a distributed system the transmission of signals is implemented as a command to a slave server. This will be explained and shown in more detail in section 5.

3.3 Shared data

In the above text it was described which communication facilities VIPWOB supports when there is a direct link between two modules (point to point) or the sender knows the receiver. In the following it is described which facilities VIPWOB has in order to support the case where sender and receiver are unware of each other.

As explained the master server maintains a simple database with a set of variables with associated values. This database is simple in the sense that variables are untyped, that is there is only one type of variables, strings. No structure of any kind is supported.

The database is used heavily by the entire system. Every definition of modules, servers, channels etc. is described in this database.

All though the database itself is simple there are, besides the standard operations like assigning values and retrieving variables, some more advanced access features. The access features are:

Set, get and delete Variables may of course, be created with or with out a value and they may be retrieved from the database. There is no difference between creating a variable and changing its value. If the variable does not exist when an attempt is made to change the value, it will be created. Finally variables may deleted.

Matching variables It is possible to get or delete a set of variables which matches a pattern. This way a set of variables with common parts in their name *or value* may be retrieved.

Locking Variables may be "locked", that is so called "destructive access" is only allowed by the module which has locked the variable. Destructive access includes changing the value and deleting the variable. Of course it is only the module which has locked the variable which can unlock it. When other modules tries to change the value or delete the variable, they will receive an error reply.

Notify on access When a module uses this feature it may receive a message (or another module may receive it.) whenever someone accesses a variable which matches a pattern. The content of the message, where it should be sent and how it should be sent is defined by the module which sets up the "notify on access" request. Any number of such setups may be present in the master server. An operation on a variable may match several "notify on access" setups in which case each of them is activated. The content of the message may include the name of the variable, the value of the variable or nothing. The message may be sent as a reply to the setup command or it may be sent as a request message. It is also possible to specify that the message should only be sent in the case that somebody deletes the variables, changes it or performs some other operation (or set of operations) on the variable. However the "notify on access" is disabled when variables are retrieved from the database using the "matching" operation described above.

Two examples of using the feature are

1. Instead of having the "options" specified on the command line, a module may have its options defined as variables in the database. Thus when somebody changes one of the variables or options the module gets a message containing the variable and the new value.
2. The feature may also be used as a synchronization mechanism. In this case a module sets up a variable and when another module accesses the variable

the module gets notified. In this case the variable and the "notify on access" behaves like an event mechanism.

4 Configuration files

Defining a startup configuration, that is specifying the modules and channels, is done by means of a text file.

In the textfile a number of "objects" are defined. These "objects" are modules, channels and slave servers. Channels and slave servers are optional; channels may be created during runtime and slave servers are defined automatically when needed. In fact the minimum configuration consists of a single module definition.

4.1 Modules

The skeleton for a module definition is shown in figure 4. A module definition

```
module <name> {
    prog = <name-of-executable>
    options = <string>
    dir = <directory-name>
    host = [<user-name>@]<host-name>
    user = <user-name>
    slave = <slave-name>
    addr = <major>.<minor>
    env = <string>
    envadd = [on|off]
}
```

Figure 4: Skeleton for module definitions

is thus a named "object" of type *module*. The "statements" within the body of the module definition are:

prog Defines the path of the executable program. This statement **must** be present. All of the following statements are optional.

options This is the string which should passed on to the module (program) when it is launched.

dir Defines the directory that the module should have as "current directory".

host and user These two fields defines the host and user (if different from the user on the local machine)

slave Defines the name of the slave server which should control the module.

addr Defines the address which should be assigned to the module.

env More than one of these statements may be present in the module definition. They defines environment variables which should be included in the environment of the module.

envadd This boolean specifies, if the value is *on*, that the module should inherit the environment from the slave server which controls it.

4.2 Channels

The skeleton for a channel definition is shown in figure 5. Naming a channel is

```
channel [<name>] {
    <link definition>
    type = inet|unix|file
}
```

Figure 5: Skeleton for channel definitions

optional. If no name is given, a unique name is generated automatically. The channel name is only used to destinguish the channel definitions. The link definition specifies the two endpoints of the channel. There are four basic types of link definitions:

<mod1>:<fd1> -> <mod2>:<fd2>
In this case the output from filedescriptor <fd1> of the module <mod1> is routed to filedescriptor <fd2> of module <mod2>.

<mod1>:<fd1> <-> <mod2>:<fd2>
Here a bidirectional channel between filedescriptor <fd1> in module <mod1> and filedescriptor <fd2> in module <mod2> is defined.

<mod1>:<fd1> -> <file>
In this case the output from filedescriptor <fd1> of module <mod1> is routed to the file specified by the name <file>.

<mod1>:<fd1> <- <file>
In this case the input to filedescriptor <fd1> of module <mod1> is take from the file specified by the name <file>.

Of course

<mod1>:<fd1> -> <file>

may be written as

<file> <- <mod1>:<fd1>

or

<mod1>:<fd1> -> <mod2>:<fd2>

as

<mod2>:<fd2> <- <mod1>:<fd1>

Note that the filedescriptor is optional. If no filedescriptor is given, one is chosen automatically. It is, however, not simple for the module to figure out which filedescriptors are assigned to which channels.

The *type* statement defines the type of the channel. It is optional and may either specified in the link definition (file links) or during runtime. Examples are provided in section 5.

4.3 Slave servers

The skeleton for a slave server definition is shown in figure 6. A slave server

```
slave <name> {
    prog = <name-of-executable>
    host = [<user-name>@]<host-name>
    user = <user-name>
    env = <string>
    envadd = [on|off]
    modenv = <string>
    rsh = <string>
    dir = <directory-name>
    addr = <major>.0
    autoquit = [on|off]
}
```

Figure 6: Skeleton for slave server definitions

definition is thus equivalent to a module definition with one exception: none of the statements **must** be used but at least one should be provided. The “statements” within the body of the slave server definition are:

prog, dir, host, user, addr and env Similar to the description for modules.

envadd This boolean specifies, if the value is *on*, that every module launched by the slave server should inherit the environment from the slave server.

modenv More than one of these statements may be present in the slave server definition. They define environment variables which should be included in the environment of the modules which are launched by this slave server.

rsh Defines a program which is capable of launching the slave server on the (possible) remote host. Note that it is also used to launch slave servers that runs on the local host.

autoquit Autoquit is normally *on*. It indicates that once the slave server has launched at least one module, the next time it sees that no modules are running it terminates automatically. This has the effect that when all modules has terminated, the entire system shutdown automatically.

4.4 Parsing

The configuration file is sent through a two-way parser. Thus statements and definitions may appear in any order. This parser automatically includes all nescessary information, thus if addresses are undefined, they are assigned, slave server definitions are created as needed, etc.

The output of this parser is a set of commands to the shell. Most of them are, however, messages that should be sent to the master server, the slave servers, and the modules.

5 Examples

This section shows and describes some examples of using VIPWOB. Examples of configuration descriptions and how to use the module programmers interface are shown.

In this section it is assumed that the names of the executables of the shell, the master server and the slave server are vwshell, vwmaster and vwslave, respectively.

5.1 A simple example

This section describes a pipe line example and it could thus also have been programmed in the standard shell. The example is used to get an idea of how VIPWOB may be used.

Assume that we want to acquire an image and extract lines from that image. The process of doing so could be done in three steps: acquire an image, find edges in the image and convert edges to lines. Assume that the program which acquires an image is named *grab*, the edge finder is called *deriche* and the line finder is named *linker*. These are the names of the executables. The flowchart of this system is shown in figure 7.

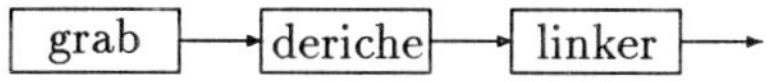

Figure 7: A simple example.

To tell VIPWOB how to launch the programs and how the programs communicate, a configuration file must be given to VIPWOB. For this simple example the configuration file is straight forward as shown in figure 8. The description of the

```
module image { prog = grab }
module edges { prog = deriche }
module lines { prog = linker }

channel { image:1 -> edges:0 }
channel { edges:1 -> lines:0 }
channel { lines:1 -> "output" }
```

Figure 8: Configuration file for the simple example.

modules is straight forward but some comments on the channels are required.

A filedescriptor is a number identifying a specific data channel. In standard Unix three filedescriptors are special: 0, 1 and 2. These filedescriptors are used for standard input, standard output, and standard error output, respectively. In VIPWOB any number of "standard" channels are allowed. The description of a channel in VIPWOB is based on the endpoints of the channel. Thus when a description of a channel looks like

image:1 -> edges:0

it implies that the standard output of the module *image* should be connected to the standard input of the module *edges.* Any other descriptors could be used as in

image:3 -> edges:0

but then the module *image* would not be as general as before since this is specific to VIPWOB and not to Unix.

Suppose a system is lauched based on this configuration file. This is done by typing "vwshell *configuration-file*". What happens now is that *vwshell* launches the master server, *vwmaster. Vwmaster* make a copy of itself to handle error I/O from the slave servers and the modules. This process is usually referred to as the master I/O process.

In this case there is a single slave server as no remote hosts are specified and no explicit definition of a slave server is provided in the configuration file.

The shell instructs the master server to launch a slave server which is given the address 1.0. Then the shell instructs this slave server to launch the modules one by one and finally the shell instructs the modules to open the channels. The opening of channels is actually a dialog between the two modules which should open a channel between them and none of the server knows about this.

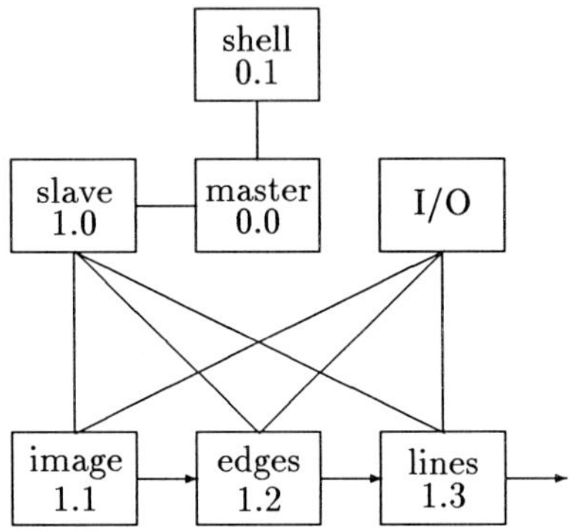

Figure 9: Configuration when everything is running.

The modules are assigned addresses in the order they are launched. In this case they are launched *image*, *edges*, and *lines*. This means that *image* gets the address 1.1, *edges* 1.2, and *lines* 1.3, respectively.

When everything is up and running the configuration with the channels is as shown in figure 9.

Note that each module has a link to the slave server and a link to the master I/O process. Also note that the master I/O process has no address, i.e. it is not a VIPWOB process and thus it is not possible to send messages to it.

5.2 An extended example

The example described next uses a feedback loop to estimate the movements of a robot. In a few words, the procedure is:

1. Set start and goal positions for the robot.

2. Acquire an image.

3. Extract lines from this image.

4. Determine robot position based on a comparison of the extracted lines and lines from a model of the world.

5. Move the robot a fraction of the difference between the current position and the goal position.

6. Jump to step 2.

A flowchart of the procedure is shown in figure 10. Note that there is a feedback

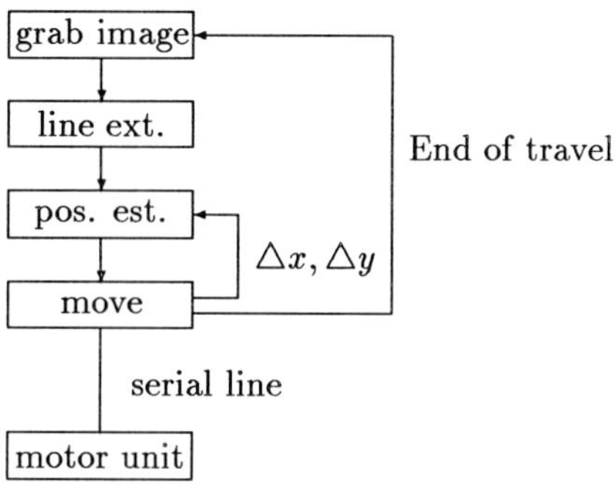

Figure 10: Movement feedback example.

from the program which does the actual movement back to the position estimator. This feedback is the actual movement and is necessary for the position estimator to track lines from one image to the next. Another thing which should be noticed is the channel between the movement program and the motor control unit. This is a serial line, not a channel inside VIPWOB. Hence it does not appear in the configuration file. Furthermore the "channel" which in the figure is called "End of travel" is not a real channel, it is more like a semaphore as discussed later.

Configuration file

The configuration file is shown in figure 11. A number of things should be noticed in the configuration file.

First of all, multiple hosts are introduced. The module *grab* runs on a specific host named *tuc* while all other modules runs on the same host as the VIPWOB shell.

Secondly this configuration file shows both the normal way of defining channel endpoints, by file-descriptor numbers and symbolic file-descriptors like "stdout".

Thirdly notice the use of the file-descriptor 3 of the module *posest*. This is a non-standard file-descriptor and the program robest must be programmed to use it, i.e. it must know about it's usage.

The "End of travel" flag

To indicate that the module *grab* may acquire a new image, the modules *move* and *grab* uses a special command of the master server. This is the **notify** command.

```
module grab
{ prog = sbhips2
  host = tuc }

module lines
{ prog = deriche2 }

module posest
{ prog = robest }

module move
{ prog = moverobot }

channel { grab:stdout -> lines:0 }
channel { lines:1 -> posest:0 }
channel { posest:1 -> move:0 }
channel { move:1 -> posest:3 }
```

Figure 11: Configuration file for feedback example.

The principle is that the module *grab* monitors a variable in the master server's database and the module *move* "changes" it. Each time the module *move* accesses this variable, the module *grab* gets notified.

To do this, assume the variable is named **GRABNEXT**. First of all the variable must be created. The module *move* does this. The code for creating (or assigning a value to a variable) can be written like

```
#include "vipwob.h"

int CreateSemaphore() {
  MessageHeader *mesg;
  byte *data;
  char cmd[30];
  int id,ret,f,l;

  strcpy(cmd,"set GRABNEXT");
  f = MESGFLAG_ACK;
  l = strlen(cmd) + 1;
  id = VWSendMessage(0,0,f,l,cmd);
  VWWaitForReply(id,&mesg,&data);
  if (mesg->flag & MESGFLAG_ERROR)
```

```
    ret = -1;
  else
    ret = 0;
  VWFree(mesg);
  VWFree(data);
  return ret;
}
.
if (CreateSemaphore()) {
  VWWriteErr("Semaphore failed\n");
  exit(1);
}
.
```

The function **VWSendMessage**() is the function used to send a message to another module. In this case the master server as the address specifier, the first two parameters, is 0.0. The third parameter defines which flags should be set in the message header. In this case an acknowledge is required from the master server. This is to ensure that the master server always sends a reply regardless of whether an error occurred or not. This is the recommended way of communicating with *any* server.

The function **VWWaitForReply**() just waits for the reply of a specific message. The last two parameters of **VWWaitForReply**() are pointers to locations where pointers to the message header and the message body should be stored. This is the normal way of receiving message from the VIPWOB library. Notice that this space is allocated by means of **malloc**() and it is the programmers responsibility to free this space after use.

Finally the function **VWWriteErr**() which acts like **fprintf(stderr,...)** except that data is transferred directly to the master I/O process in one chunk.

Using the master server command **notify** the module *grab* can subscribe to changes of the variable. This is may be done like

```
#include "vipwob.h"
.
int StartNotify() {
  MessageHeader *mesg;
  byte *data;
  char cmd[30];
  int id,l,f;

  strcpy(cmd,"notify g GRABNEXT -f r");
  f = MESGFLAG_ACK;
  l = strlen(cmd) + 1;
  id = VWSendMessage(0,0,f,l,cmd);
```

```
  VWWaitForReply(id,&mesg,&data);
  if (mesg->flag & MESGFLAG_ERROR)
    id = -1;
  VWFree(mesg);
  VWFree(data);
  return id;
}
.
if ((notifyid = StartNotify()) == -1)
{
  VWWriteErr("Notify failed\n");
  exit(1);
}
.
```

Notice again the use of MESGFLAG_ACK. The usage of **notify** in this case states: When somebody tries to read the variable called **GRABNEXT**, return an empty reply message to the source of the command (the module *grab*). The id of the reply message will be equal to the id of the message which contained the **notify** command. The reply message will be empty since no message string is supplied.

Now the module *grab* must wait for another module to access the variable.

```
void WaitForGo(notifyid)
int notifyid;
{
  MessageHeader *mesg;
  byte *data;

  VWWaitForReply(notifyid,&mesg,&data);
  VWFree(mesg);
  VWFree(data);
}
.
/* the variable notifyid
   was defined earlier */
for(;;) {
  WaitForGo(notifyid);
/* grab an image and write
   it to standard output */
}
```

The last part is to get the module *mode* to access the variable. This code is shown below.

```
void GrabNext() {
  MessageHeader *mesg;
  byte *data;
  char cmd[30];
  int id,l,f;

  strcpy(cmd,"get GRABNEXT");
  l = strlen(cmd) + 1;
  f = 0;
  id = VWSendMessage(0,0,f,l,cmd);
  VWWaitForReply(id,&mesg,&data);
  VWFree(mesg);
  VWFree(data);
}
.
  GrabNext();
.
```

Notice that when the module *move* tries to access the variable, there is always a reply, regardless of whether an error occurred or not. If the flag MESGFLAG_WARN is set somebody has erased the variable. To prevent this, a small modification to the function CreateSemaphore could be made. The new version is described below.

```
int CreateSemaphore() {
  MessageHeader *mesg;
  byte *data;
  char cmd[30];
  int id,ret,f,l;

  strcpy(cmd,"lock GRABNEXT");
  l = strlen(cmd) + 1;
  f = MESGFLAG_ACK;
  id = VWSendMessage(0,0,f,l,cmd);
  VWWaitForReply(id,&mesg,&data);
  if (mesg->flag & MESGFLAG_ERROR)
    ret = -1;
  else
    ret = 0;
  VWFree(mesg);
  VWFree(data);
  return ret;
}
```

Instead of using the **set** command, the **lock** command is used. This command creates the variable **GRABNEXT** with no value assigned to it and prevents all other modules from changing or erasing the variable. However, other modules may still examine the variable, thus it still serves as a "flag".

5.3 Sending and receiving messages

The previous section introduced the art of sending and receiving messages. This section describes the subject with more examples.

Non-waiting replies

In the previous example **VWSendMessage**() followed by **VWWaitForReply**() was used to send a message and wait for the reply. Suppose the time waiting for the reply should be used more efficiently. One way to do this is:

```
l = strlen(cmd) + 1;
f = MESGFLAG_ACK;
id = VWSendMessage(0,0,f,l,cmd);
while(!VWGetReplyFromQueue(id,&mesg,&d))
{
  /* something else */
}
/* process the message */
VWFree(mesg);
VWFree(data);
```

The function **VWGetReplyFromQueue**() returns 0 as long as the reply hasn't arrived yet. Once the message arrives, **VWGetReplyFromQueue**() reads the message and returns 1.

Multiple destinations

More than one module may receive a specific message. A broadcast message is the type of message sent when one or both of the address specifiers of **VWSendMessage**() is -1. As a single example suppose *all* modules should receive a specific message. Then the address specifier should look like

```
  f = MESGFLAG_ACK;
  l = strlen(cmd) + 1;
  id = VWSendMessage(0,-1,f,l,cmd);
```

Notice that the MESGFLAG_ACK is set. Thus all modules should reply to this message, sending a number of reply messages with the *same* id.

Message interception

It is not always convenient to wait for a message explicitly. Therefore VIPWOB supplies a way of using an "interrupt" handler when a message arrives. The principle is that when a slave server has a pending message to a module, it signals the module using a userdefined signal. This signal is caught by the VIPWOB library and if the module has supplied an interception function, this function is called. Note that this happens as soon as the message has been read, i.e. *before* it is put into the message input queue. The module now has the opportunity to react on the message immediately or just put it into the message input queue and react later. A skeleton is shown below.

```
.
int MessageHandler(mesg,data)
MessageHeader *mesg;
byte *data;
{
  /* check the message:
    should we react now ? */

  /* yes, do some processing */
    return 1;

  /* no, do it later */
    return 0;
}
.
  /* turn on message interception */
  VWIntercept(MessageHandler,SIGUSR2);
.
  /* turn off message interception */
  VWIntercept(NULL,0);
.
```

Thus if the intercept functions returns 0, the message is put into the input queue and may be extracted by other functions. Otherwise it is assumed that it did respond to the message and the message is discarded afterwards.

5.4 Signaling

By using the appropriate slave server it is possible for a module to send a signal to another module. Such a message should be sent to the slave server controlling the module to which the signal should be sent. There are currently no support in the slave servers to route such messages. However the master server has a command, **resolve** which is used to find the address of a named module or server. This way a module may find the address of another module, strip the minor address number (setting it to zero) and send a signal message to that slave server.

In a previous example a variable in the master server's database was used to indicate that the next image should be acquired. Instead this could be acheived using signals. Assume that the module *grab* reacts upon the signal SIGUSR1, then the code in module *grab* could be written as

```
#include <signal.h>

int SignalReceived = 0;

void GotSignal() {
  SignalReceived = 1;
  signal(SIGUSR1,GotSignal);
}
.
  signal(SIGUSR1,GotSignal);
  for(;;) {
    SignalReceived = 0;
    while(SignalReceived == 0)
      pause(); /* take a nap */
    /* grab an image and write
       it to standard output */
  }
.
```

Notice that the code is independent of VIPWOB ! In the previous example the module *grab* was running on a host named *tuc* and all other modules was running on the same host as the VIPWOB shell. Assume that this is two different hosts and thus there are two slave servers, one for the module *grab* and one for all other modules. Furthermore assume that the slave server which controls the module *grab* has the address 1.0 and the other slave server has the address 2.0. Now for the module *move* to send a signal to the module *grab* it must send a message to the slave server controlling *grab*. This could be written as

```
#include "vipwob.h"

void GrabNext() {
  MessageHeader *mesg;
  byte *data;
  char cmd[30];
  int id,f,l;

  sprintf(cmd,"signal grab %d",SIGUSR1);
  f = MESGFLAG_ACK;
```

```
  l = strlen(cmd) + 1;
  id = VWSendMessage(1,0,f,l,cmd);
  VWWaitForReply(id,&mesg,&data);
  VWFree(mesg);
  VWFree(data);
}
.
  GrabNext();
.
```

This assumes that the two machines has the same value for SIGUSR1. An alternative is to use named signals but this requires a signal environment in the receiving slave server.

There are in addition several commands for accessing the database of variables in the master server, but the facilities mentioned above are the main facilities for module to module communication.

6 Experimental Evaluation

6.1 Benchmarking

In order to evaluate the transfer rate or rather the overhead imposed by VIP-WOB on the transfer rate, a number of tests has been performed. These tests include:

- Transfer rate of point to point channels as a function of data size between modules on the same host.
- Transfer rate of point to point channels as a function of data size between modules on different hosts.
- Transfer rate of messages as a function of data size between modules on the same host.
- Transfer rate of messages as a function of data size between modules on different hosts.
- "Transfer rate" of the *notify on access* feature.
- "Transfer rate" of signals between modules with common slave server.

To compare the times measured from these tests with the corresponding times of a "dedicated" system, two comparability tests has been made:

- The transfer rate of an Internet socket between two processes running on two different hosts.
- The transfer rate of a Unix socket between two processes.

A short comment to the times shown in the table: The measured times are processor time, not real time[2]. This implies two things:

Table 1: Timing results for transferring data between modules, both for "raw" transfer (not using VIPWOB) and in a VIPWOB system. All times are in milliseconds, average of 5000 tests.

Setup	Init	Time / Kb
Transfer between modules on the same host	0.20	0.021
Transfer between modules on different hosts	0.26	0.20
Transfer between modules on the same host, using VIPWOB	0.28	0.027
Transfer between modules on different hosts, using VIPWOB	0.34	0.19
Message parsing between two modules, common host	1.5	0.15
Message parsing between two modules, different hosts	1.6	0.14
Accessing a global variable on the same host	1.6	
Accessing a global variable on another host	1.7	
Signalling a module on the same host	1.6	
Signalling a module on another host	1.7	

1. The time only represents the time used by a single process. This is especially important to note for message timing. Ex. when two modules are sending messages to each other through a slave server, the time only represent the time spent in a single module. *It does not include the time spent in the slave server or the other module.*

2. The time does not include the time spent to transfer data through the communication media. *Ex. the time spent to transfer data on the Internet network is not included.*

Before commenting on the figures, it should be noted that when these tests was performed the standard deviation of the time measurements was approximately 10 to 50 % of the mean value. This indicates that the measurements should not be taken too literally.

The main results of these performance figures are:

1. There is a slight increase in initiation time when using VIPWOB. This time is used to check for incoming messages. Otherwise no performance degradation.

2. The performance of messages is comparable whether a messages is destined for a local module or a remote module.

[2]For the technical minded the measured time is the sum of the usertime and systemtime returned by the system function **times**(2)

3. Messages has a relative large initiation time but same transfer time as direct transfer of data.

As commented above, the time measurements for messages does not include the time spent in any server or another module. Thus transfer time for messages should in fact be multiplied by 2 or 5, depending on whether the modules has common slave server or not.

7 Summary

VIPWOB has been designed to allow construction of computer vision systems and attention has especially been paid to design of facilities that allow use of VIPWOB ranging from being a simple execution environment to detailed control of the interaction between modules.

VIPWOB has been implemented on a wide range of UNIX platforms, including HPUX series 300 and 700 running HPUX 7.0 and HPUX 8.0, Motorola MVME147 and MVME167 running SVR5r3 and SVR5r4, SGI Personal Iris and Indigo running IRIS 3.3.2 and IRIX 4.0.5 and Sun Sparc running SunOS 4.0.1.

Porting VIPWOB to other platforms may typically be done in less than a day. Currently the only requirement is that the platform runs UNIX and BSD sockets are available.

VIPWOB is public domain software but copyrighted according to the GNU software license. It is available through anonymous ftp from danablue.vision.auc.dk in the directory *pub/vipwob*

This release contains the complete software and documentation. The software is autoconfiguring, thus installation is trivial.

Users Guide and Reference Manual are also available in PostScript format on danablue.

8 Acknowledgement

The VIPWOB project was initiated in 1988 as part of the MOBS programme. A large number of people have participated in the project either as designers or debuggers. Prof. Erik Granum, Prof. Knut Conradsen and Prof. Peter Johansen conveived the project, while Carsten Trant initially headed the effort. Bjarne Ersbøll, Jens Michael Carstensen, Mikael Kjærulf, Mads Nielsen, and Peter L. Petersen has through out the project participated in the design of the system. Peter Melsen programmed the first two releases. The help of these people are greatfully acknowledged. In addition is the funding of Niels O. S. Kirkeby from the Danish Technical research Council and the funding of Henrik I Christensen from the EEC acknowledged.

References

1. J. Rasure, *The Khoros System*, in Scientific Visualisation, p. 106, ed. R. A. Earnshaw and N. Wiseman, Springer Verlag, 1989.

2. L. Olsson, *Module Network Tool*, Tech note CVAP-TN11, Computational Vision and Active Perception Laboratory, Royal Institute of Technology, Stockholm, 1992.

3. Sillicon Graphics Inc., *The Explorer System*, Tech Report, Mountain View, CA, 1991

4. Advanced Visual Systems Inc., *The AVS System 4.0*, Tech Report, Los Angeles, CA, 1989

5. N. O. S. Kirkeby and H. I. Christensen, *The VIPWOB 4.0 Documentation Suite*, LIA-Tech Report 93-02, Institute of Electronic Systems, Aalborg University, 1993.

6. B. Kernighan and D. Ritchie, *The C Programming Language*, Prentice-Hall Inc., 1986.

7. S. Kristensen, *Implementation of HELIX facilities under VIPWOB 4.0*, LIA-Tech Report 93-10, Institute of Electronic Systems, Aalborg University, 1993.

MODULE NETWORK TOOL

LARS OLSSON
Computational Vision and Active Perception Laboratory (CVAP), Royal Institute of Technology (KTH), S-10044 Stockholm, Sweden
Email: lolsson@bion.kth.se

ABSTRACT

Computational networks are software systems made up of a set of interconnected computing elements called *modules*. Modules are concurrently executing processes with a fixed number of input/output *ports*. A computational network can be thought of as a graph where nodes are modules and arcs are port to port connections. The arcs carry streams of data items that are transformed at the nodes. This computational model fits well with many image processing applications where streams of images and image features are handled by a variety of operators and filters. Furthermore, the reactiveness of modules makes computational networks particularly well suited to active vision systems where high level strategies must coexist with reactive low level behaviors.

Tools for the construction of computational networks usually divide development into two distinct phases: Modules are usually programmed in a standard programming language, whereas networks are constructed using dedicated network specification languages. This paper describes a tool that supports construction of computational networks running in UNIX environments. Modules correspond to processes, and port to port connections are based on sockets. Modules can be programmed in C++. A launcher program sets up a network according to the directives in a network specification.

1. Introduction

Building large scale experimental Computer Vision systems usually involves considerable software integration efforts. Such systems are typically made up of many separate pieces of software which must somehow be put together. A closely related problem is to share software among individual researchers and between research groups. Many low and intermediate level processing algorithms may be used in widely different high-level systems, and it therefore is natural to ask how code reuse can be supported.

What is needed is a software integration tool which allows users to rapidly assemble the low and intermediate level parts from existing pieces of software, so that attention may be turned to the implementation of the new, high level components. The resulting systems should be highly modular, allowing one piece of software to be easily replaced by a functionally equivalent one. There are many algorithms that solve the same task, and it often interesting to see what happens to the system as a whole when different versions of an algorithm are used. Modular design also makes it possible to replace software which controls specialized hardware devices with programs simulating the actions of these devices, thus allowing systems to be ported to sites where the specialized hardware is not available. Even work on early processing algorithms may benefit from the use of software integration tools. Much may be learnt about the quality of the output of an early processing algorithm if it is embedded in a larger system.

The simplest software integration mechanism is the subroutine library. This approach gives maximum flexibility but it does not allow rapid prototyping. Subroutines must be called from a main program which must be compiled and linked, before the program can be executed and tested. UNIX shell pipe-lines are less time-consuming for the user to handle, but are less flexible. Furthermore, various properties of computer vision systems make the subroutine library and UNIX shell pipe-line unsuitable for integration purposes. It is natural to view computer vision systems as collections of connected filters. Data streams pass through the system where outputs of early processing filters are fed into down-stream filters. In the simplest case the system consists of a simple pipe-line of filters where data flows in one direction. However, it is also possible to have a system composed of pipe-lines with multiple branches, where filters in different branches may execute in parallel. This is where the UNIX pipe-lines approach fails. Furthermore, research directions such as *Animate Vision*[2] bring with them requirements for feedback mechanisms: data driven reflexive processes must coexist with top-down driven visual tasks, and results of computations performed in high-level components may be fed back to lower-level components such as the controller of a camera head.

While it is in principle possible to implement such systems with the subroutine library approach, in practice too much is required of the programmer who must explicitly handle process forking, synchronization, and interprocess communication. Instead, a number of software integration tools have been designed to support the construction of systems which may be referred to as *computational networks*. The basic software component is usually called a *module* and a computational network consist of a set of interconnected, independently executing modules which communicate via message passing.

This paper will describe one computational network tool, called *Module Network Tool* (MNT). The paper is organized as follows: Section 2 discusses various properties of computational network tools, Section 3 describes some existing tools, and MNT is described in detail in the remaining sections.

2. Properties of Computational Networks and Tools

Over the years both commercial vendors and university research groups have developed various systems that can be described as computational network tools. Some of these systems are devoted solely to network construction, whereas others have wider scope, and network construction is only one of many capabilities.

A number of existing systems will be described in Section 3. This section sets the stage for these descriptions by listing a number of important characteristics of computational network tools. Basically the list highlights a number of design considerations that may be addressed in different ways. These considerations have impact on both the tools themselves, and the corresponding computational networks. No claims are made that the list presented below is complete in any way:

- *Two-phase development*: Modules are typically written in a standard programming language and implementation support comes in the form of a library and an Application Programmer Interface (API) for module construction. Networks are usually started by a *launcher* program which places modules on processors according to the directions in a network specification. This specification may be a text file containing statements in some network specification language, or it may be generated with a graphical network editor where networks are built interactively.

  It is far from obvious that system development should be divided into two main phases: module construction and network construction. For instance, in the ADA programming language[7] a process is an independently executing piece of code with a well-defined interface, but it is also a basic element of the programming language. A process may be defined as an ADA type so that processes are started when variables of that type are created. Thus the distinction between processes and the systems they are part of becomes blurred.

  The main reason development is divided into two distinct phases is that it facilitates rapid prototyping. The idea is that the second activity, that of network construction, should be relatively easy and straight-forward. While module implementation may require considerable programming skills, network construction should involve little more than listing the modules and specifying how they should communicate. If a graphical network editor is available this will be particularly easy. As the library of modules grows it becomes progressively easier to construct large scale systems.

- *Message passing mechanism*: There are at least two different message passing models. Traditionally, low-level image processing systems have been implemented using the *point-to-point connection* paradigm: internally a module sends and receives messages via a set of local ports, which the launcher will connect to ports of other modules. In contrast, high-level reasoning systems have usually been structured as a set of processes which communicate via a shared *database*. Each process may post data items to the database, or request notification of database events.

- *Supported machine architectures*: During network specification developers place modules on processors. Tools differ in how *portable* modules are. Some tools tie modules to one specific processor architecture, some allow entire networks to be ported to different architectures, and a few tools allow developers to build mixed-architecture networks, where different modules belonging to *one* network execute on different kinds of processors.

- *Data representation*: If modules are to communicate, they must agree on a communication protocol. The *data representation* problem is an important one which can be approached in different ways. One approach is to provide a standard but flexible set of data structures on which all communication must be based. The opposite approach is to provide only the basic byte stream transfer mechanisms and let the modules themselves decide on the communication protocol.

- *Parameter specification mechanisms*: Most algorithms in the computer vision field are controlled by a number of parameters. Software integration tools must provide mechanisms which allow the user to give these parameters different values, preferably while the system is executing. Two approaches are possible: either the mechanisms are built into the software integration tool itself, or they are provided by specialized parameter control modules (that may be modified and extended by users).

3. Some Existing Tools

Computational Networks are useful in a number of application domains, and quite a few network construction tools have been implemented. This section describes a few systems that have received attention in the vision community.

The software development environment for Transputers[8] fits into the point-to-point connection approach: Development can be divided into a logical phase where programmers deal with *processes* which communicate via *channels*, and a physical phase where processes are mapped to *processors* and channels are mapped to processor-to-processor *links*. These mappings are specified in a configuration file processed by a server program which places processes on processors and allocates a link for each channel.

The VIPWOB[9] system provides distribution capabilities in UNIX environments. A VIPWOB application consists of a set of UNIX processes which have been started on different processors by a distributed server program according to the directions in a configuration file. These processes send and receive messages by writing to and reading from ordinary file descriptors. The VIPWOB runtime system takes over the UNIX system calls `open()`, `close()`, `read()`, and `write()`, and associates output file descriptors of one process with input file descriptors of another process.

The ANIT[4] system supports mixed-architecture networks. It is possible to have networks where some modules execute as UNIX processes, and other modules execute as Transputer processes. ANIT has a C-based module construction API. Programmers are required to provide C-functions to do initialization, input handling, output handling, and main processing. There are utility functions to create and manipulate input and output *wires*, as well functions to define and use algorithm parameters called *control vectors*. For every module in a network, the ANIT run-time system generates an X window with interaction glyphs for each control vector defined by the module.

In AVS,[1] a module is a procedure written in C or Fortran, with one procedure parameter for each module port. Procedures are compiled into a form that can be handled by the AVS *flow executive*, which controls the execution of AVS networks by calling module procedures when input port values change. The flow executive is integrated with the *network editor* which allows networks be specified interactively. AVS ports are *typed*, and the network editor makes sure that port types match when users connect modules. There is a distinction between data input ports and parameter ports in that parameter ports may be associated with pop-up windows where parameter values can be set interactively. AVS provides a set of predefined data types with one port type for each data type, as well as a large library of modules which communicate using these types. AVS runs on UNIX systems with X Windows[a].

Another UNIX-based tool is KHOROS[11] in which programs called *vroutines* (or *xvroutines* if they have a graphical user interface) are developed separately, and then combined into complete systems by the *Cantata* program.

Examples of systems based on shared databases are the *Schema* system,[5] and *Core Knowledge Structure*.[13]

4. Module Network Tool

The tool described in this paper, *Module Network Tool* (MNT) has been designed to be a small, extendible, and policy-free computational network tool. More specifically:

- MNT consists of a very small core package which provides only the basic module creation mechanisms, and a set of extension packages.
- Applications developed with the core package do not depend on X Windows in any way. Support for modules with X Window interfaces is provided in an extension package.
- The core package provides a simple static module parameterization mechanism. Support for run-time control of parameters is provided in an extension package.
- The core package contains a launcher API from which arbitrary launcher programs may be built. A simple text based launcher based on this API is also available, and a graphical network editor is under construction.
- MNT does not restrict data that can be sent between module to a fixed set of data types. The core package contains a port type implementation API

[a] X Windows is a trademark of M.I.T.

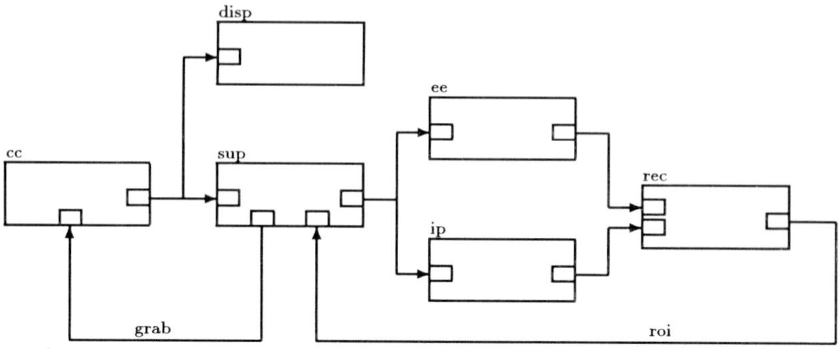

Fig. 1. A 2D recognition network. A supervisor module *sup* sends image grab requests to a camera control module *cc*, which sends images to the supervisor and a display module *disp*. The supervisor cuts out a region of interest which is sent to an edge extraction module *ee* and to an interest point detection module *ip*. Their respective outputs are used by a recognition module *rec* to generate object hypotheses. The recognition module passes a region of interest request to the supervisor when it wants a new image.

which enables programmers to create port types handling arbitrary data transfer protocols, as well as two predefined port types.

- Image processing and rendering support is provided in an extension package.

MNT can be used to construct computational networks based on point-to-point connections. An MNT module is a process with a fixed (but arbitrary) number of bidirectional[b] and typed *ports*. A module consumes data items arriving on input ports and produces derived data items on output ports. It may also have a state which evolves as input streams are processed. Modules are event driven and synchronization within a network is achieved solely through data transfer between modules: a module either performs some background computation or sits idle and waits for the arrival of new input items. Whenever a new input item arrives on a port, the module wakes up, fetches the item, processes it, updates its state if it has one, writes derived data to some of its output ports, and goes back to sleep.

It is helpful to think of an MNT network as a graph where nodes are modules and arcs are port-to-port connections (Fig. 1). The MNT launcher ensures that only ports with matching types are connected.

Most of the MNT development effort has gone into the C++[6] based module construction API, which comes in the form of a number of C++ base classes which

[b] An MNT port may be used for both input and output. In what follows, an input port is simply a port which is not used for output, and an output port is a port for which no input is expected.

may be subclassed[c] by the application programmer. The C++ subclassing mechanism may be used to implement modules, new port types, specialized launchers, special event detectors (e.g. X-events), and daemon handlers.

5. Modules

An MNT module is implemented as a separately compiled program, which is executed as a UNIX process. The launcher may launch multiple instances[d] of a module, in which case each instance executes as a separate UNIX process. When a module is invoked, the launcher passes certain arguments in `argv` to the `main` function of the module program:

```
int main(int argc, char *argv[])
{
  MntScheduler scheduler;                       // scheduler object
  MntServer    server(argc,argv,scheduler); // server object
  Edge         ee(scheduler,server);            // module object

  scheduler.Dispatch();                         // give control to scheduler
}
```

The `main` function creates three objects, a *scheduler*, a *server*, and a *module* object. The server object uses the information in `argv` to set up a connection to the launcher program. When this connection has been established, the server can set up port-to-port connections to other modules under the direction of the launcher. Once the start-up phase is over, the server is responsible for the detection and processing of data arriving on ports.

The application specific details are placed in a module object[e] (called `ee` here). This object contains definitions of the ports used by the module, definitions of the state maintained, and code for processing data arriving on ports.

The module is controlled by a *scheduler object.* Once control has been given to the scheduler through a call to its `Dispatch` function, the scheduler, server, and module objects cooperate in the following fashion (Fig. 2) :

- The scheduler repeatedly executes a cycle in which it tries to divide processing time between port input processing and background computations.
- At the beginning of every cycle the server is given control so that it can check the ports to see if there is any new input to be processed.

[c] An extension, or specialization of a class is called a subclass or a derived class.

[d] Strictly speaking, a module network consists of a set of *module instances*. However, when there is no risk of ambiguity, the term *module* will be used instead.

[e] Since every module program has exactly one module object together with the scheduler and server object shared by all module programs, it makes sense to be somewhat inexact and use the same term "module" when referring to the module program, the module class, and the module object.

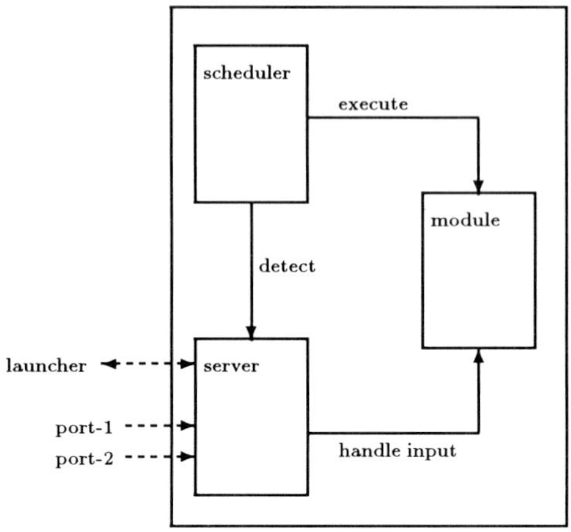

Fig. 2. The scheduler controls the flow of execution in a module. It regularly calls the server so that it can detect new port input events. Whenever this occurs the server calls the module object to handle the new input data. When no input data is available the scheduler may call the module object to execute some background code.

- If the server finds that there is new input available on a port, it gives control to the module object so that the new data can be processed.
- If the server did not detect new input data on any of the ports, the scheduler may give control to the module object so that it can perform some background computations.

Application programmers may change this scheduling strategy by redefining the `Dispatch` function in a scheduler subclass.

Programming a module amounts to creating a module class which contains the application specific data and code. Module classes should be derived from the module base class (`MntModule`) which defines data and functions common to all modules, most notably the input handling function `HandleInput`, and the background processing function `Execute`.

The module subclass declaration usually contains member objects which represent the ports and parameters (called *resources* in MNT, see Section 7) used by the module, as well as the members representing the state of the module. The input handling function is redefined if port input data is expected, and the background processing function is redefined if the module intends to perform background computations:

```
class Edge : public MntModule {
protected:
```

(a)

```
Edge::Edge(MntScheduler &sh,MntServer &sv)
:MntModule("edge-extract",sh,sv),
 in_port("in",this,this),
 out_port("out",0,this),
 threshold("threshold",this),
 gray(0),edge(0)
{
}
```

(b)

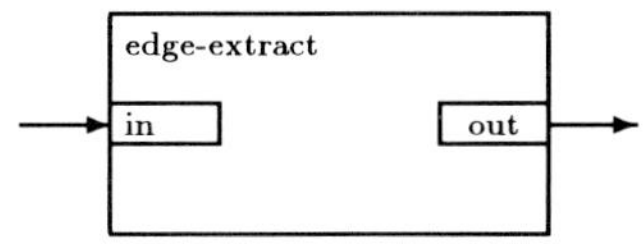

Fig. 3. The module constructor function may serve as an external specification of a module. (a) shows the code for the constructor function and (b) shows a pictorial representation of the external interfaces of the module. The module name *edge-extract* is specified in the call to the module base class constructor. The names of the ports and the resource are specified in the corresponding constructor calls.

```
  MntXorPort        in;
  MntXorPort        out;
  MntResourceInt    threshold;
  Image            *gray;
  Image            *edge;
public:
                    Edge(MntScheduler &sh,MntServer &sv);
  void              HandleInput(MntPort *port);
};
```

The `Edge` class represents a filter module that processes gray-level images and produces edge images on its output port.

Module initialization takes place in the module class constructor function which is called automatically when the module object is created (Fig. 3). The module base class constructor is executed first, then the individual member objects are initialized, and finally the module subclass initialization is performed. During module base class initialization a confirmation message is sent back to the launcher so that it can check that the launched program contains code for the expected module. A message is also sent to the launcher whenever a port object is initialized. This message contains all the information required by the launcher to set up port-to-port connections. Resource objects are initialized with values provided by the launcher.

The module class declaration and constructor function together define the structure of the module. The class declaration defines the data and code used by the module, and the constructor usually provides the external names (the names used by the launcher) for the module, its ports, and its resources. To find out what a module does it should be sufficient to inspect the module class declaration. The constructor function contains the information required to include the module in a module network.

6. Ports

When a port has pending data the input handling function of the module is usually called. The only argument to this function is a pointer to the port object which represents the port. Given this pointer the input handling function can call read functions provided by the port object to fetch and process the pending input data.

There are different port classes handling different data transfer protocols. MNT provides two port classes: class `MntBinaryPort` supports transfer of binary buffers, and class `MntXorPort` supports transfer of objects or clusters of objects according to the XOR[14] protocol. Every port class is subclassed from the abstract port base class `MntPort`.

Each port class provides its own set of service functions to send and receive data. For instance, the XOR port class has a `Write` function to send objects via ports, and a `Read` function to receive objects from ports[f]. In the example below it is assumed that there is a function `compute_edge_image_from()` which computes an edge image given a gray-level image:

```
void Edge::HandleInput(MntPort *port)
{
  if (port == &in) {
    delete gray;
    delete edge;
    gray = Image::cast(in.Read());
    edge = compute_edge_image_from(gray);
    out_port.Write(edge);
  }
  else {
    Trace("Unexpected input for port %s\n",port->Name());
    port->Discard();
  }
}
```

The `Edge` module expects images on the port represented by port object `in`. If the input handler is called for another port the module writes a warning message to the log file and discards the pending data.

The `Read` function returns a pointer to the next input object. Every class derived from class Xor is automatically provided with a `cast()` function that casts the returned pointer to the correct type, or terminates the module if the object is not an instance of that type[g]. The caller becomes the owner of the object and has to make sure that it is deleted when it is no longer needed. In this example the module does not delete its images immediately, but waits until the next input image arrives.

---

[f] Objects must be instances of classes derived from the `Xor` class.

[g] An alternative function `cast0()` does not terminate the process, but returns a `NULL` pointer when the object is not of the expected type.

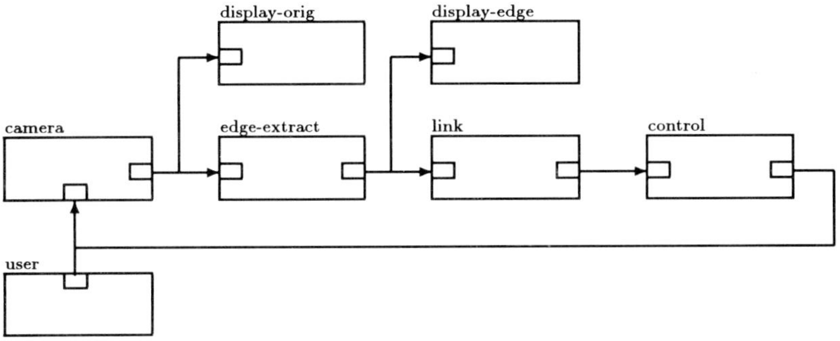

Fig. 4. A camera module sends images to an edge extraction module which sends computed edge images to an edge linking module. This module sends linked edges to a control module, which in turn sends image grab requests to the camera module. A module called *user* allows users to interactively send commands to the camera module during execution. The outputs of the camera and edge extraction modules are also passed to image rendering modules.

When the launcher sets up a network (Fig. 4) of modules, it establishes port-to-port connections between the modules. Normally, an output port of one module is connected to the input port of one other module. However, an output port of one module may be connected to input ports of any number of other modules, In that case the sending port object is responsible for copying the data output to every connected module. Similarly, an input port of a module may be connected to an arbitrary number of sending modules, in which case the input port object is responsible for merging the separate input data streams into a single stream before making it available to the receiving module.

Port objects are also responsible for synchronization. If a sending module is considerably faster than the receiving module it is connected to, the output port of the sending module will sooner or later suspend execution of the sender until the receiver has processed its backlog of input data[h].

At run-time, knowledge about how the module is connected to other modules is stored in the port objects, and a module may receive and send data via ports without knowing how it is actually connected to other modules. This means that modules are fairly independent of the modules they are connected to. Modules must of course agree on what kind of data to transfer but they do not need to know each other by name. It is this independence that makes modules reusable software components that can be multiply instantiated in networks.

[h] This behaviour may lead to deadlocks in networks with feedback loops.

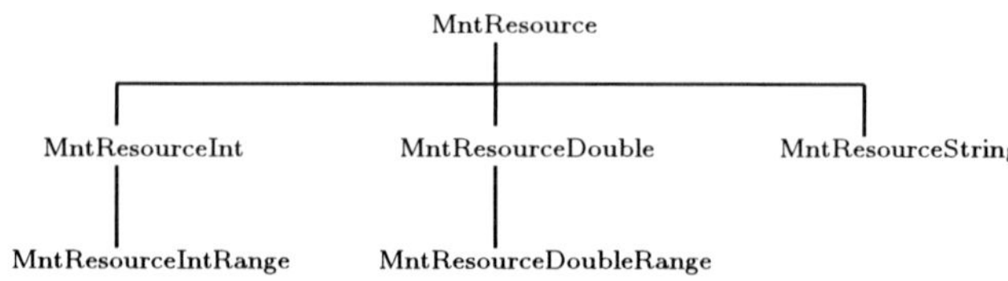

Fig. 5. Resource classes are arranged in a hierarchy which contains resource classes to handle integer, floating point, and string values.

7. Resources

Many algorithms are parameterized, and a computational network tool must allow users to set such parameters from outside modules, even during execution if desired. MNT parameters are called *resources* and are represented by resource objects. There are different resource classes for different kind of values (integers, floats, and strings), and each resource class provides a set of functions for setting and getting resource values (Fig. 5). All resource classes are derived from a common resource base class and it is possible to extend the resource class hierarchy with new subclasses to handle new resource types.

There are at least three ways that users might want to set resource values:

- Most resources have a value that does not need to be changed while the network is executing. Furthermore, once good values have been found for the resources one would like to use these values every time the network is executed. A natural place to specify values for such resources is in the network specification.

- Sometimes it might be desirable to change resource values interactively as the network is executing. The idea is that the user should be able to "tune" the network while it is running by changing resource values and inspecting the corresponding changes in module outputs.

- Finally, it should be possible to let value of a resource be be controlled by the outputs from another module.

Resource value control mechanisms are encapsulated in resource *controller* objects. A resource object may be associated with any number of controller objects, each of which may set the value of the resource. Different resource controller classes handle the different ways that resources can be set. MNT provides three classes to handle the three cases described above:

MntModule The module base class is partially derived from the resource controller base class. It uses the values provided by the launcher to set values for resources. The launcher in turn gets these values from the network specification.

MntResourcePort An `MntXorPort` subclass which maintains a set of resource objects, that it tries to update whenever an new XOR object arrives.

MntResourceRepository A subclass of `MntResourcePort` which supports interactive control of resource values. When a resource object is added to the set of maintained resources, a definition of the added object is sent via the `MntResourcePort` port to a specialized *resource repository module*. This module accepts resource definitions from client modules, and creates a user-interaction glyph for each received resource. These glyphs enable the user to modify the value of any resource. When a value is modified, it is sent back to the repository port of the client module, where it is used to update the appropriate resource object.

These classes are all derived from a resource controller base class, and it is possible to extend the controller class hierarchy with new subclasses to add support for new ways to control the values of resources.

Since a resource object may be controlled by several resource controller objects of different classes it is possible for a resource object to have its value set in several ways. For instance, the `Edge` module of Section 5 defines an integer resource object, called **`threshold`**, which is initialized with the module object as the associated resource controller object, This means that the value is fetched from the network specification used by the launcher. Support for dynamic run-time control of this resource can be added in a subclass to `Edge`, where the resource object is placed under the control of a repository port:

```
Edge2::Edge2(MntScheduler &sh,MntServer &sv)
:Edge(sh,sv),                                  // initialize superclass
 repository_port("repository",this,this)       // repository port
{
  threshold.AddResourceSource(repository_port); // associate
}
```

Fig. 6 shows graphically how the **`threshold`** resource is controlled by two resource controllers. When the user changes a resource value, the repository module sends the new value to the `Edge2` module, which causes the input handling function of this module to be called:

```
void Edge2::HandleInput(MntPort *port)
{
  if (port == &repository_port) {
    repository_port.Check();
    if (threshold.Changed())
      Trace("threshold=%d\n",threshold->Int());
  }
  else
    Edge::HandleInput();
}
```

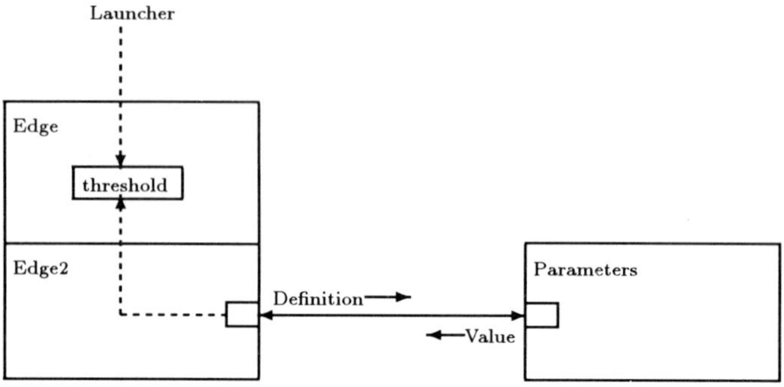

Fig. 6. The resource called **threshold** is under the control of the launcher *and* a repository module called *Parameters*.

A repository port may maintain several associated resource objects, and when a new value arrives, it must update the appropriate object. This is done by the function **Check** which must be called in the input handling function of the module. This should be followed by a test to see which of the associated resource objects was updated. The function **Changed** returns non-zero if the resource object has been updated since the previous call to **Changed**. It is then up to the programmer to decide what to do when a resource value has changed. One approach is to do nothing and to simply use the new value when new input data arrives and is processed. Alternatively, the input data can processed again as soon as a resource value is changed. This of course requires that the module keeps the input data after it has been processed.

8. Background Processing

When no data is arriving on any of the ports of a module, the scheduler object may try to call the background processing function of the module, allowing it to perform background computations. The function, which is called **Execute**, is defined in the module base class, and it should be redefined in module subclasses that intend to perform background computations:

```
class EdgeGroup : public Edge {
protected:
  Grouper    grouper;
  MntXorPort group_port;
public:
             EdgeGroup(MntScheduler &sch,MntServer &srv);
  void       HandleInput(MntPort *port);
  void       Execute(void);
```

```
};
```

The above example shows how the edge extraction module of Section 5 can be extended to search for sets of edge groupings in the background. The input handling function has been redefined since the background computation must be initialized at some point, and it is natural to do this once a new edge image has been computed:

```
void EdgeGroup::HandleInput(MntPort *port)
{
  Edge::HandleInput(port); // normal processing
  grouper.Use(edge);       // initiate search
  Enable();                // enable background processing
}
```

The input handling function first calls the input handling function of its superclass which fetches a gray-level image and computes and sends an edge image, just as before. The actual search for sets of edge groups is performed by a special `grouper` object (of assumed type `Grouper`). The input handling function initiates the search for edge groupings by calling the `Use` function of this class which informs the grouper object which edge image to search in. Finally, background processing is *enabled*: for reasons of efficiency, a module object may have background processing enabled or disabled (the default). The module base class defines two functions `Enable` and `Disable` which enable and disable background processing respectively. The background processing function will be called if there is no input on any of the ports *and* background processing is enabled:

```
void EdgeGroup::Execute()
{
  Groupings *groupings = group.Search(); // get some new groupings
  if (groupings)                         // if there were any new groupings
    group_port.Write(groupings);         // send them
  else                                   // no new groupings found
    Disable();                           // don't try anymore
  delete groupings;
}
```

The `grouper` object function `Search` spends a limited amount of time in searching for sets of edge groups. It returns a set of groupings (of assumed type `Groupings`) which is sent via a special port called `group_port`. If the search fails to find any new groupings it returns zero. Since subsequent calls to this function with the same edge data would also fail, background processing is immediately disabled.

As long as no port input data arrives, the background processing function will be called repeatedly. Each time the `grouper` object searches for some groupings and returns quickly. It keeps track of where it postponed its search so that it can be resumed in subsequent calls to codeSearch. The search for sets of edge groupings is thus spread out over several calls to the background computation function. This allows the module to respond quickly to new port input data, since the server object is given the opportunity to try to detect new input data between calls to `Execute`.

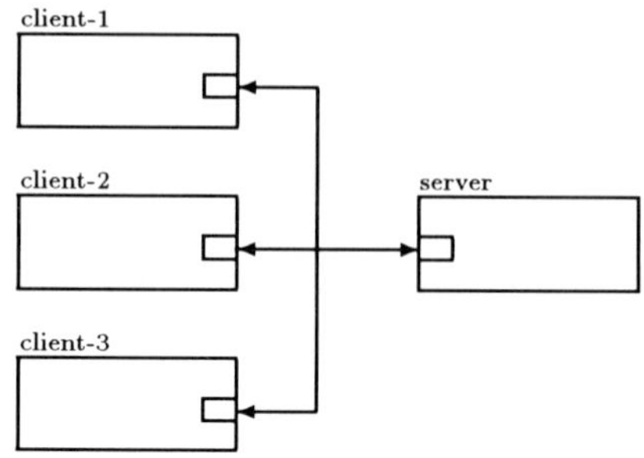

Fig. 7. A data base server module connected to three client modules.

9. Port Connection Descriptors

Sometimes the mechanism which causes port objects to automatically send data to multiple modules is inappropriate. Consider a database server module with one *client* port to which an arbitrary number of client modules may be connected (Fig. 7).

The input handler function of such as server module could be coded as follows:

```
void Server::HandleInput(MntPort *port)
{
  if (port == &client_port) {
    Request *request = Request::cast(client_port.Read());
    Reply *reply     = compute_reply(request);
    client_port.Write(reply);
    delete reply;
    delete request;
  }
}
```

This example is straight-forward: a request is received, and a reply is computed and sent back. However, now the hidden duplication mechanism causes the reply to be sent to *all* connected client modules, not only the client which sent the request.

The use of *connection descriptors* allows programmers to avoid this unacceptable behaviour. Every port object maintains a list of connection descriptors, where each descriptor represents an actual port-to-port connection. Send and receive functions usually permit connection descriptors to be specified, thereby giving the programmer more knowledge about the origins of input data, and control over the destination of output data. For instance, the XOR port class provides an alternative version of the `Read`[i] function which returns the connection descriptor together with the fetched

[i] C++ functions may be overloaded, that is several functions taking different arguments may share the same name.

object. The descriptor indirectly[j] identifies the module which actually sent the object. When a connection descriptor is specified in a call to the `Write` function the object is sent only to the module identified by the descriptor[k]. Using connection descriptors the server module can be rewritten as shown below:

```
void Server::HandleInput(MntPort *port)
{
  if (port == &client) {
    MntXorPortConnection *client_conn;
    Request *request = Request::cast(client_port.Read(client_conn));
    Reply   *reply   = compute_reply_from(request);

    // send to requesting module only
    client_port.Write(reply,client_conn);

    delete reply;
    delete request;
  }
}
```

10. Port Groups

Another problem with the port mechanism, as presented so far, is that it is not possible to synchronize the arrival of data items coming from different ports. For instance, a stereo processing module would typically have two input ports, one for images taken by the left camera, and one for images from the right camera. Since these images have to be processed together one would like to fetch both of them at the same time. But since the input handling function is called as soon as input arrives on a port, it is generally not safe to assume that data is available on two different ports at the same time. What is needed is a mechanism which causes the call to the input handling function to be postponed until data is available on both ports.

To this end, MNT provides *port group* objects, which are special port objects that control a set of normal port objects. A port group ensures that the input handling function is not called until data is available on all controlled ports:

```
class Stereo : public MntModule {
protected:
  MntPortGroup images;    // port group
  MntXorPort   left;      // left image port
  MntXorPort   right;     // right image port
  MntXorPort   out;       // output port
```

[j] It identifies a connection between the port and a port of another module.

[k] A connection descriptor is valid for *one* port object only. Even though two ports may be connected to the same module it is not possible to specify a connection descriptor belonging to one port when calling the `Write` function of the other port.

```
public:
                Stereo(MntScheduler &sch,MntServer &srv);
  void          HandleInput(MntPort *port);
};
```

Normally, a module calls the constructors of its port objects with a reference to itself (`this`) to make the port objects call the `HandleInput` function of the module when new input data must be processed. The port group object replaces the module object as the input handler of the ports that are to be controlled:

```
Stereo::Stereo(MntScheduler &sch,MntServer &srv)
:MntModule("stereo",sch,srv)
 images(this),
 left("left",&images,this)
 right("right",&images,this),
 out("out",0,this)
{
}
```

Since the port group object itself has the module as its input handler, the programmer may be sure that data is available from every controlled port when the input handler function of the module is called:

```
void Stereo::HandleInput(MntPort *port)
{
  if (port == &images) {
    Image *left_image  = Image::cast(left.Read());
    Image *right_image = Image::cast(right.Read());
    Image *depth       = stereo_match(left_image,right_image);
    out.Write(depth);
    delete depth;
    delete left_image;
    delete right_image;
  }
}
```

11. Port Subclassing

New port types can be created as subclasses of the port base class `MntPort`. The base class deals with the multiple connections a port may have, while port subclasses deal with the actual data transfer protocol (Fig. 8). The base class maintains a list of connection descriptors, one for each module the port is connected to. A connection descriptor contains information about the module it represents, but it also contains a handle to the underlying connection, which in the current MNT implementation is a socket descriptor.

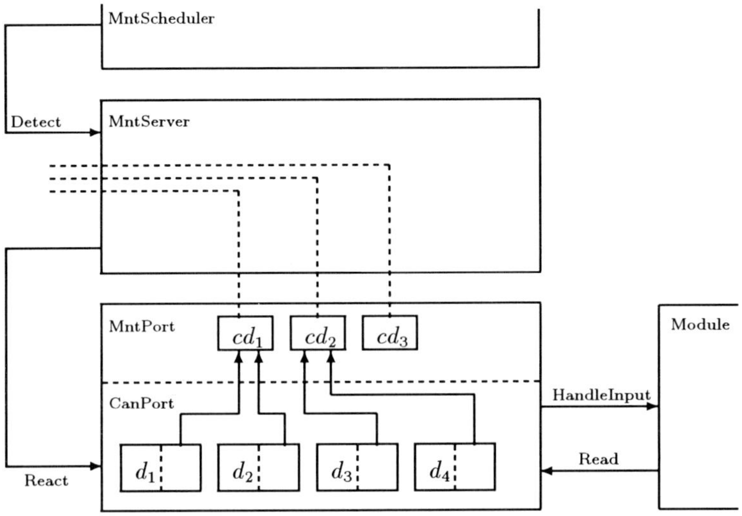

Fig. 8. A port object maintains a set of connection descriptors $\{cd_i\}$, and a queue of pending input items $\{d_j\}$. When the server detects new port input it calls the `React` function of the port to append new items to the internal queue. As long as the queue is non-empty the module object input handler is called. This function removes items from the queue by calling a `Read` function provided by the port.

A port subclass must provide functions to send data items via ports, either to every module connected to the port, or to a single module identified by a connection descriptor. Port output is relatively straight-forward: the send function may write data to only one specific connection descriptor, or it may loop through the set of connection descriptors and write the data to each one.

Similarly, the port subclass should merge data items arriving from multiple sources into a single stream, and it should provide functions which fetch items from this stream, with or without connections descriptors which identify the source of the fetched item. Port input is a bit more complicated than port output: a port must maintain a queue of pending data items, it must contain a function called `React` which is called (by the server object) whenever new data becomes available on a connection, and it must contain functions that fetch data from the queue of pending data items. The `React` function should append new data items to the end of the queue, and the fetch functions should remove data items from the front of the queue. The elements of the queue should contain a data part and a pointer to a connection descriptor which identifies the module which sent the data item.

Consider implementing a port class that supports transfer of CANDELA[12] images.

The name of the port class will be CanPort and the name of the protocol will be "candela". The protocol name is specified in the CanPort constructor:

```
CanPort::CanPort(const char *name,
                 MntPortInputHandler *inputh,
                 MntPortOwner *owner)
:MntPort(name,"candela",inputh,owner) // protocol name is "candela"
{
  can_init();                          // initialize candela system
}
```

The port class provides a function to send images, either to all connected modules, or only to the module identified by a connection descriptor:

```
void CanPort::Write(canPic pic,MntPortConnection *connection)
{
  if (connection) {
    int fd = connection->OutDescriptor();
    can_writepic(fd,pic)
  }
  else
    for (int i = 0 ; i < Connections() ; i++) {
      int fd = Connection(i)->OutDescriptor();
      can_writepic(fd,pic);
    }
}
```

The connection descriptor member function OutDescriptor() returns the underlying socket descriptor used for image output. The function can_writepic is the CANDELA image writing function. If connection is not NULL, the image is sent via that connection only. Otherwise the function loops through all connections and sends the image to each in turn.

For input, the class maintains a list of items which contain an image and a pointer to the connection descriptor from which the image was fetched:

```
class CanItem : public MntObject {
public:
  canPic            pic;
  MntPortConnection *connection;
  CanItem(canPic p,MntPortConnection *c) {pic=p; connection=c;}
};
```

The React function fetches the image from the underlying socket descriptor, then creates and appends a CanItem object to its internal queue[l]:

[l] The items list is of type MntList, and is defined in the MntPort base class. This list class can store list elements of type MntObject which is why CanItem is derived from MntObject. The MntList member function append appends an item to the end of the list.

```
void CanPort::React(MntPortConnection *c)
{
  int      f     = c->InDescriptor(); // get underlying file descriptor
  canPic   p     = can_readpic(f);    // get image
  CanItem *item  = new CanItem(p,c);  // create pair item
  items.append(item,1);               // append to internal stream
}
```

The connection descriptor member function InDescriptor() returns the underlying socket descriptor which is used to fetch the image. The can_readpic function is the CANDELA image reading function.

The class also needs to supply functions which fetch the next image from the internal queue. The first version of this function sets the argument connection to point at the corresponding connection descriptor, whereas the second version, which calls the first, discards the connection descriptor.

```
canPic CanPort::Read(MntPortConnection *&connection)
{
  MntPix   pix      = items.first();          // point to head of list
  CanItem *item     = (CanItem *)items(pix);  // extract list element
  canPic   p        = item->pic;              // get image
  connection        = item->connection;       // get connection descriptor
  items.del(pix);                             // remove list head
  return p;                                   // return image pointer
}

canPic CanPort::Read()
{
  MntPortConnection *connection;  // connection descriptor to be ignored
  return Read(connection);        // fetch image
}
```

12. The Module Scheduler - Listener Objects

The MNT scheduler maintains a list of *listener* objects which detect and process different kinds of input events. Properties common to all listeners are encapsulated in a listener base class from which specialized listeners can be derived. The server object is a specialized listener which detects input events on a set of ports, and this is usually the only listener a module has.

Modules with X Windows interfaces usually have two listener objects, the server object and an X-event detector. This detector is provided in an MNT extension package, but a simplified version will be shown here to demonstrate listener implementation.

The listener base class defines a virtual function Listen() which is called by the scheduler to allow the listener to detect and process events:

```
class XtListener : public MntListener {
private:
  XtAppContext appcon;
public:
               XtListener(XtAppContext ac,MntScheduler &scheduler);
  int          Listen(int timeout);
};
```

The `Listen()` function returns the number of processed events. If no new events are immediately available the listener may choose to go into a wait state until an event occurs or a timeout expires. The `timeout` argument specifies the length of the timeout period in milliseconds.

The X-event listener shown here detects and processes X-events. It ignores the timeout argument and returns only when there are no more events to process:

```
int XtListener::Listen(int timeout)
{
  int processed = 0;              // # of processed events
  while(XtAppPending(appcon)) {   // while more pending events
    XEvent e;
    XtAppNextEvent(appcon,&e);    // get next event
    XtDispatchEvent(&e);          // process it
    processed++;                  // count it as processed
  }
  return processed;
}
```

13. The Module Scheduler - Step Objects

In addition to listener objects the scheduler maintains a list of background processors called *step* objects. It loops through this list in a round robin fashion, giving control to *one* step object if all listeners have failed to detect input events. The scheduler gives control to a step object by calling a member function `Execute()`, defined in the step base class `MntStep`. Specialized background processors can be created by deriving subclasses from this class, and the alert reader will have noticed that the module base class inherits its `Execute` function from the step base class.

In this section, the use of step objects will be illustrated by showing how a simple daemon facility might be implemented. A module, called the daemon handler, maintains a state of some kind, and allows computational entities called daemons to monitor this state. Each daemon is programmed to check for the occurrence of some event. Client modules create and send daemons to the daemon handler which includes them in the set of actively monitoring daemons. When a daemon detects "its" event, it notifies the client module it originated from.

In the implementation below, each daemon will be a step object which will check for its condition in its `Execute` function. To be able to send a notification message to its client module, the daemon needs references to the port and port connection from which it has been fetched by the daemon handler module:

```
class Daemon : public MntStep , public Xor {
public:
  void Client(MntXorPort *port,MntXorPortConnection *connection);
  void Execute(void);
};
```

In the Execute function, the daemon would make use of the port object and the port connection descriptor passed to the Client function to send its notification message to the client module.

A daemon handler module would typically have a port to receive daemons from client modules. When a daemon arrives it will be installed as a step object through a call to the scheduler object function Step[m]:

```
void DaemonModule::HandleInput(MntPort *port)
{
  if (port == &daemon_port) {
    MntXorPortConnection *connection;
    Daemon *daemon = Daemon::cast(daemon_port.Read(connection));
    daemon->Client(daemon_port,connection); // inform about client module
    scheduler->Step(daemon);                // install as step object
    daemon->Enable();                       // activate daemon
    daemon_list.append(daemon,1);
  }
}
```

Once installed, the daemon will check for its condition whenever its Execute function is called by the scheduler.

14. The Launcher

Launcher programs are responsible for placing modules on processors, for setting up port-to-port connections between modules, and for setting resources for each module. MNT is shipped with a simple text-based launcher which reads a text file which contains statements for creating module instances, for setting up port-to-port connections, and for setting resource values for the instances.

The *module* statement declares a module instance, specifying the instance name, the module name, the name of the executable file, and an optional machine name.

[m] Objects fetched from ports must be deleted at some point. In this example, it is not possible to delete the daemon object immediately, as it must exist in order to monitor the state of the module. Since the scheduler does not assume ownership of its step objects, the module must keep the daemon objects in a list (called daemon_list here) so that they can be deleted later on. The second argument (1) to the list append function specifies that the list object should assume ownership of the object being appended.

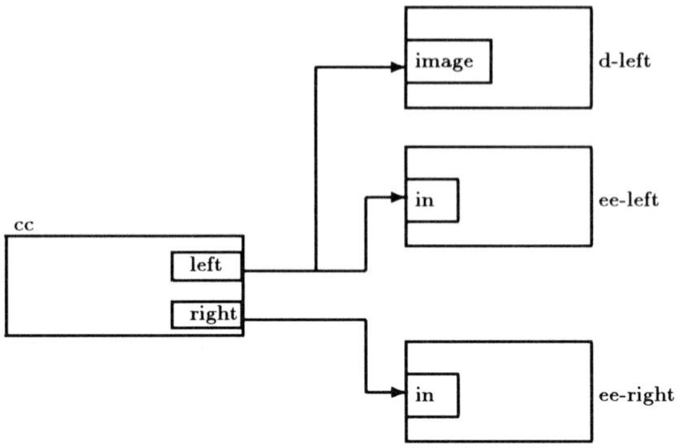

Fig. 9. A network consisting of a camera control module, two edge extraction modules, and an image display module.

For instance, the following statements list the module instances of the network shown in Fig. 9.

```
(module cc        camera-control $HOME/bin/sun4/cc chaplin.bion.kth.se)
(module ee-left   edge-extract   $HOME/bin/sun4/ee)
(module ee-right  edge-extract   $HOME/bin/sun4/ee)
(module d-left display $HOME/bin/sun4/display)
```

Port-to-port connections are specified by the *link* statement, which associates a port of one module instance with a port of another module instance.

```
(link cc left  ee-left in)
(link cc right ee-right in)
(link cc left d-left image)
```

If the edge extraction module has a threshold parameter, it can be specified as a resource as follows:

```
(resource ee-left  threshold 20)
(resource ee-right threshold 20)
```

15. Module library

MNT has to date only been used to construct a few non-trivial systems. One idea behind computational network systems is that as new systems are built, it should be possible to take the general purpose modules of these systems and add them to a module library. Over time, as the library grows, it becomes progressively easier to rapidly construct systems from ready-made components. Since MNT has not yet been extensively used, its module library is quite small. The following is a brief description and classification of modules implemented so far:

- One class of modules fetch and store data from external (to the network) sources. For instance, MNT has various modules to read and write images from/to disk, as well as X window based modules that allow users to interactively inject scalar data items into networks. The most important of these is the *parameter repository* module which allows users to control the values of resources used by other modules.

- Display modules allow users to inspect the intermediate results computed inside a network during execution. For instance, MNT has a *display* module that shows images in an X window, as well as two *progress report* modules that show the user how a computation performed by a module proceeds.

- Image processing and image analysis modules typically process incoming images and produce enhanced images or sets of features extracted from the input image. For instance, MNT has modules to compute edge maps, pyramids, and sets of interest points.

- Driver modules controls special hardware devices, or the execution of programs that are not MNT modules, thus making it possible to include such programs in a network. For instance, the MNT *system* module calls the `system()` function, thus making it possible to execute any UNIX command, or any ordinary program. The *camera control* module controls the KTH head-eye system.[10] This service module may be connected to an arbitrary number of client modules. These may send commands to set camera parameters and to grab images.

16. Discussion

MNT has been designed with the needs of active vision systems in mind. However, it is a completely general computational network tool that can be used in a variety of application domains. Great emphasis has been put on providing a convenient API and in making the tool extendible. MNT's strongest feature is its object-oriented module construction API which accounts for most of the flexibility of the tool. In summary, some of the advantages of MNT are:

- A module is implemented as a class derived from a module base class. Using the inheritance mechanism of C++ it is possible to achieve code reuse not only on the level of individual modules, but also on the level of module subclasses.

- Similarly, there is a hierarchy of Port classes that the application programmer can add to. Specialized classes can be implemented to handle specific data types and communication protocols. The standard port to `MntXorPort` provides object persistence when used together with the XOR package.

- MNT also provides a hierarchy of resource value types, and a hierarchy of resource controllers. The former specifies what kinds of parameters a module may have, and the latter controls the mechanism by which parameter values are set. Once again, the application programmer may add to the hierarchies if the functionality of the standard classes are inadequate.

- MNT provides a launcher base class from which different launcher programs can be constructed. The MNT distribution comes with a simple text-based launcher program.

- The high level of abstraction makes MNT portable. Presently, modules are implemented as UNIX processes, and port connections are based on sockets. In future releases, modules could be mapped to processes on a Transputer, or tasks in a co-routine system, and ports could be mapped to Transputer links, or queues in a shared memory system.

- The connection capabilities are quite flexible. The output of one module may be connected to the inputs of an arbitrary number of modules, and the input of a module may be connected to the outputs of any number of modules. Since ports allow two-way communication it is possible to implement server modules accepting an arbitrary number of clients.

Some weaknesses of MNT are:

- In the present implementation, every module is a UNIX process. This allocation may be quite wasteful, since UNIX processes are expensive resources.

- The current MNT implementation can neither detect nor handle deadlock situations.

- The library of existing modules is quite small.

- There are some important system designs that cannot easily be supported by MNT. For instance, there is currently no direct support for construction of systems based on the *subsumption architecture*.[3]

17. Summary

MNT is a small, modular and extendible computational network tool. While it has been specifically designed to allow for the construction of active vision systems, it may be used to construct stream-based system with feed-back mechanisms and reactive behaviors in other application domains.

Currently, there is an MNT implementation for Solaris 1.0[n]. Modules execute as UNIX processes which may be distributed over a network of host computers. Port-to-port connections are based on UNIX sockets. Individual modules may be programmed in the C++ language.

In performance tests it was found that the using MNT rather than ordinary UNIX processes connected with pipes brought about a five percent decrease in throughput.

MNT is distributed under the GNU General Public License, and is available by anonymous ftp from `ftp@bion.kth.se`.

References

1. Advanced Visual Systems. *AVS Developers Guide*, March 1992.
2. D. H. Ballard. Animate vision. *Artificial Intelligence*, 48:57–86, 1991.
3. R.A. Brooks and J. Connel. Asynchronous distributed control system for a mobile robot. In *SPIE Conference on Mobile Robots*, 1986.
4. S. Cornell. *ANIT Programmers Guide.* AI Vision Research Unit, University of Sheffield, May 1993.
5. B. A. Draper, R. T. Collins, J. Brolio, A. R. Hanson, and E. M. Riseman. The schema system. *International Journal of Computer Vision*, 2:209–250, 1989.
6. M.A. Ellis and B. Stroustrup. *The Annotated C++ Reference Manual.* Addison-Wesley, 1990.
7. J. Ichbiah. Preliminary ada reference manual and rationale for the design of the ada programming language. *SIGPLAN Notices*, 14(6), 1979.
8. INMOS. *ANSI C Toolset User Manual*, August 1990.
9. N. Kirkeby and H. Christensen. *VIPWOB 3.1 Description and Reference Manual.* Laboratory of Image Analysis, University of Aalborg, April 1992.
10. K. Pahlavan and J.-O. Eklundh. A head-eye system — analysis and design. *Computer Vision, Graphics, and Image Processing: Image Understanding*, 56(1):41–56, 1992. Also TRITA-NA-P9208.
11. J. Rasure, D. Argiro, T. Sauer, and C. Williams. A visual language and software development environment for image processing. *International Journal of Imaging Systems and Technology*, 2:183–199, 1990.

[n] Solaris is a trademark of Sun Microsystems

12. M. Rendahl, J. Gårding, and T. Lindeberg. Candela – an overview. Tech. Note CVAP-TN07, October 1990. (Also in Proc. Aalborg University Vision Workshop, Nov 22, 1990).

13. T. M. Strat. *Natural Object Recognition*, chapter Appendix A, pages 149–164. Springer, 1992.

14. H. Winroth. External object representation. Technical Report TRITA-NA-P9128, November 1991.

OBJECT-ORIENTED COMMUNICATION IN IMAGE PROCESSING SYSTEMS

HARALD WINROTH
Computational Vision and Active Perception Laboratory (CVAP),
Royal Institute of Technology (KTH), S-10044 Stockholm, Sweden
Email: harald@bion.kth.se

ABSTRACT

Most image processing systems contain a set of predefined data structures which represent images, edges, regions etc. However, there is usually little support for user-defined data types or for extensions of existing types. Furthermore, it is often difficult to store customized data structures in files, or to send them between different processes.

In this article, an object-oriented communication scheme based on persistent objects is presented. It will be demonstrated how such a mechanism can be incorporated into several existing image processing systems and what the advantages of the approach are. In this context, some general principles of object-oriented design and how they can be applied to so called computational networks will also be discussed.

Keywords: image processing, object-oriented, persistent object, communication, module network.

1. Introduction

Many image processing systems are most naturally described in terms of streams and filters, where each filter reads data from one or more input streams and produces derived data on its outputs. In its simplest form, such a stream-based system can be implemented as a simple UNIX shell pipeline, consisting of separately compiled programs. However, for streams with branches and feedback loops, such a simplistic approach is inadequate. More support for combining filters and other computational agents (e.g. databases and blackboards) into complex systems is needed.

A lot of filters, especially those involved in the early processing stages, can be used in many image processing algorithms. To simplify code reuse we need to support the construction of such standard software components from which complete systems can be built. Moreover, it is often necessary to spread out computations over several processors, e.g. the workstations in a local area network. The system developer should be able to utilize such coarse-grained parallelism, without having to worry about the details of process communication and synchronization.

Several distributed stream-based systems that meet these demands exist today. KHOROS is an image processing system developed at University of New Mexico.[1] AVS is a commercial visualization system from Advanced Visual Systems Inc.[2] VIPWOB (VIsion Programmers WOrkBench) was developed at the Laboratory of Image

Analysis at Aalborg University in Denmark, and is distributed freely.[3] MNT (Module Network Tool) is another research software package developed within the CVAP group at the Royal Institute of Technology in Stockholm, Sweden.[4]

In all of these systems, networks consist of more or less independently executing filters, called *modules*. These modules are compiled into executable programs, which can be combined interactively by the user into a complete network, see Fig. 1.

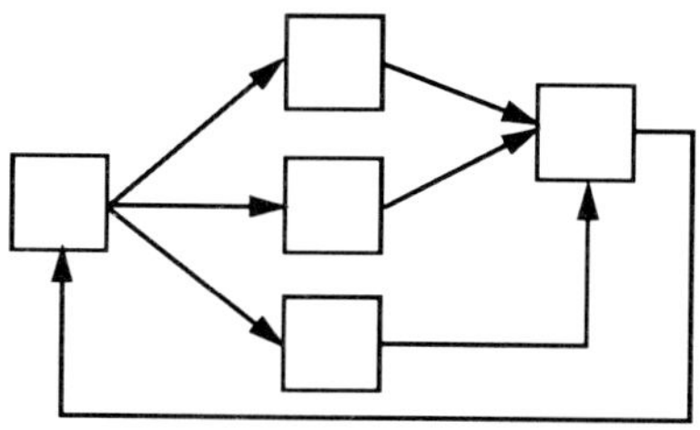

Fig. 1. A sample computational network with branches and a feedback loop.

In some systems, e.g. AVS, several modules may share a single operating system process. In general, however, each module executes as a separate process in its own address space. Therefore, data have to be transferred between processes using the mechanisms available in the operating system, e.g. temporary files, sockets, and shared memory segments. The protocol used for this module-to-module communication may be predefined,[2] or may be left entirely to the application (module) programmer.[3] In the former case, the programmer can usually choose from a set of standard data structures that can be sent between modules in a network transparent way. In systems not committed to a particular protocol, there is typically some kind of "generalized file descriptor" available, which represents a low-level communication channel (a byte stream) and which can be used as a base for high-level communication layers.

The choice of module communication protocol has far-reaching consequences for the overall system organization. One of the major concerns in this context is to determine the amount of information to encode into a particular representation. The amount of detail in the low-level data structures depends heavily on the demands of high-level computations. For example, in computer vision applications a simple geometrical object such as an edge can be interpreted in different ways depending on its origin: it can be an intensity discontinuity, the border of a textured region, or the co-linear alignment of other features. Some algorithms will only need the geometrical properties (orientation, length etc.) that all edges share, while more specialized algorithms might need, for instance, the contrast of intensity edges or the list of co-linear features forming an apparent edge. As the system evolves to handle new tasks, high-level processing steps often need more detailed information from the low-level layers. The introduction of new, extended data types usually has consequences for the whole pipeline and may require substantial modifications to the source code.

In this paper, we will try to identify such communication-related problems which often occur in computational networks and discuss how they could be solved. In particular, it will be shown how all the computational network systems mentioned above could benefit from an *object-oriented* communication scheme. Rather than presenting a complete experimental environment, an object-oriented communication mechanism that can be used in existing systems will be described.

2. Selecting a Network Communication Protocol

If two modules are to communicate with each other, they must speak the same language, i.e. use the same communication protocol. When selecting a protocol, one must carefully consider its consequences for system flexibility, code reuse, reliability, and portability. Some of the systems mentioned in the previous section choose specific protocols, while others provide only a basic, policy-free communication mechanism on which application-specific protocols can be layered.

With these systems, the users can build new applications by combining existing modules into new network configurations. The purpose of using precompiled modules is twofold. First, complex systems can be assembled quickly and interactively (rapid prototyping). Second, a significant amount of code reuse is achievable at the level of modules. Obviously, the more general the modules are, the higher degree of code reuse can be achieved. This is in turn determined by two factors: the extent to which the computations performed by a module can be controlled and modified without recoding and recompiling it, and the flexibility of the communication protocols used by the module.

Obviously, it is desirable to have a flexible protocol, since it allows any members chosen from a large set of compatible modules to be connected. This can be achieved by transferring only reasonably general data structures, such as scalars, vectors and matrices, while avoiding very specific structures with limited applicability. This is the approach taken in e.g. AVS, where the most important data type is a generalized multi-dimensional vector. However, the modules will then often be compatible only on a very low, physical level: a module may be able to read a data item, but will not be able to interpret its contents correctly, or in the way the sender intended. Hence, the modules are not necessarily semantically compatible, even though they use the same (physical) data format.

The data compatibility problem is closely related to that of *data integrity*. With only very general and unconstrained data structures available, modules have to rely on conventions and ad-hoc rules for representing data. In particular, such structures make it very hard to enforce data invariants throughout the network. For example, assume that we have an image smoothing module whose input and output ports use some kind of image encoding protocol, see Fig. 2a. Now suppose that a user wants to attach histograms to some of the images processed in the network. The first problem will of course be to fit the histogram into the existing data format. Some image encoding schemes are actually flexible enough to accommodate a histogram, for example in a free-format label stored with the images.[5] A more serious problem, though, is that the existing image processing modules will not be aware of the histogram and its semantics. Consequently, they cannot be expected to update it properly.

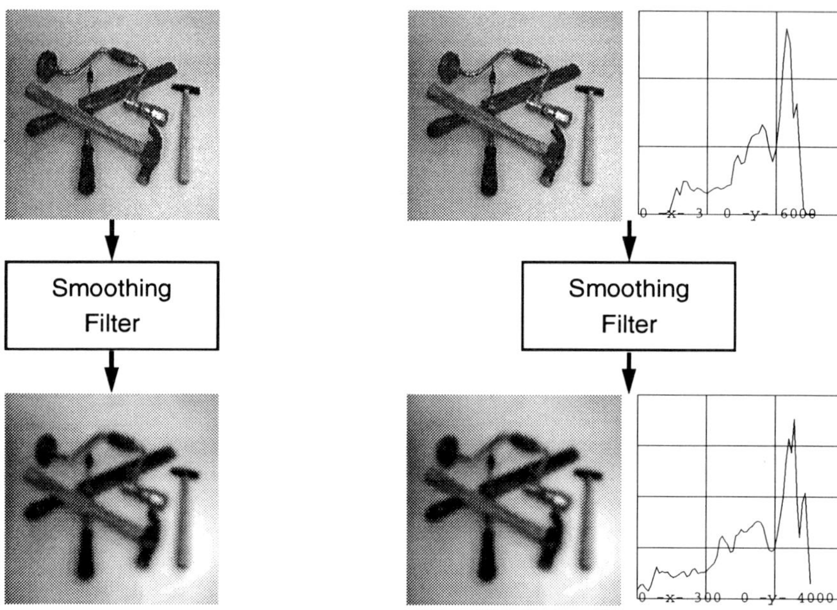

(a) (b)

Fig. 2. (a) A smoothing module designed to process ordinary 2-D images. (b) If the images are extended with histograms, the externally exported access functions of the new image objects must make sure that the histogram parts are always consistent with the pixel parts.

Therefore, all modules involved have to be modified when new data types are introduced, to avoid inconsistencies in the new structures. Although this would in principle be possible, there are severe practical problems with such an approach. First of all, the module source code may not be available—a common situation when the modules are part of a commercial system. Even if the source code is available, modules shared among users will probably be stored in protected libraries, and can only be modified if the users make their own private copies. Since users defining a new data type would probably need a substantial number of standard modules, all of which must be modified to handle the new type, this strategy would diminish the advantages gained with computational networks, such as rapid prototyping and code reuse. Furthermore, it would result in a large number of similar modules, which would only differ slightly in the way they handle certain data types. On the other hand, if every shared module were to be updated each time a new data type were introduced by some user, the source code could very quickly get bulky and error-prone.

An alternative approach is to associate the new code with the objects themselves. The modules can then be implemented with *generic operations* only, while the objects provide the code for the actual operations performed. This requires some kind of *late binding*, where the exact meaning of an operation is not determined until the actual object type is known to the module.

A consistent use of late binding allows existing modules to manipulate new, extended data types safely, without violating the integrity of the data. In the image-with-histogram example above, each (destructive) pixel access function or filtering operation could be bound at run-time to functions that would also update the histogram, or at least mark it as invalid, see Fig. 2b.

It should be pointed out that well-defined protocols can also significantly increase software portability: if all network communication strictly adhere to a general protocol defined in terms of objects, most of the code implementing the modules and the objects they manipulate can be reused in any other network systems or stand-alone programs that are able to support the protocol. The possibility of using the same code in several systems does not seem to have been exploited to any larger extent before. (However, both KHOROS and VIPWOB do provide mechanisms for using modules also as regular programs in UNIX shell pipelines.)

These considerations suggest the use of a communication protocol based on typed objects. To summarize, the aim of the approach would be

- to allow users to define new data structures for network communication at any time,
- to provide mechanisms enforcing the integrity of such data structures, and
- to support extension and specialization of all available data structures, in a way that preserves backward compatibility with existing modules, even if these have already been compiled.

Furthermore, from a practical point of view, the question of portability is equally important. In particular, one could expect

- new implementation languages for modules to be introduced,
- existing application programs to be re-implemented as network modules, and
- modules to be moved between different environments and computational network systems.

This implies that the protocol should be programming language independent in order to allow modules written in different languages to communicate, and that it should be possible to implement it with rudimentary support from the underlying communication channels (e.g. byte streams). Finally, it can be anticipated that several versions of a particular data structure have to be maintained simultaneously. Therefore, version control facilities should be an integral part of the protocol.

There are several ways of providing the necessary mechanism, both at the abstract programming level and at the operating system level. This will be discussed in Section 4. First, however, some relevant object-oriented concepts will be discussed in the next section.

3. Applying Object-Oriented Concepts

The previous section touched upon the concept of *data encapsulation.* In object-based systems, the *object* is the basic unit of encapsulation: it consists of an internal data structure representing its state, which can only be accessed through a set of *methods*, see Fig. 3. The methods thus constitute the external interface of the object.

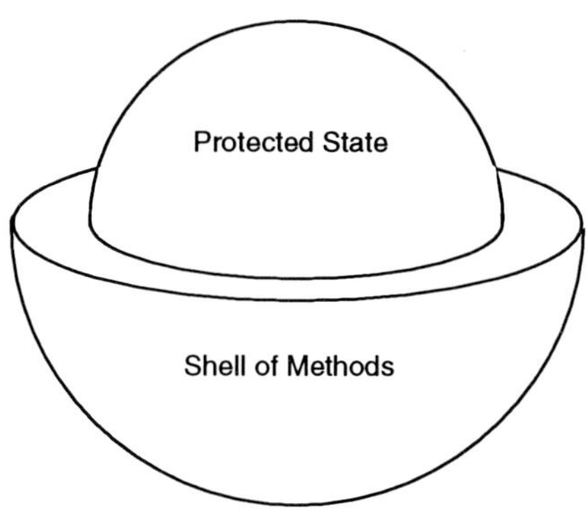

Fig. 3. The state of an object is represented by a nucleus of data structures which are accessible only through exported methods.

As mentioned before, the most important aspect of data encapsulation is its ability to ensure data integrity: by restricting all external access to the internal data structures of an object, the object state can be guaranteed to be consistent at all times. This will in turn ensure that the object always behaves in a predictable way, according to the specification of its external interface.

Data encapsulation also gives the implementor of an object the freedom to change the actual internal data representation, without affecting old application program code. Moreover, the external interface may support multiple views of the same internal state. For example, a line can be described by its slant and its distance from the origin, or alternatively, by the coefficients of its equation (see Fig. 4).

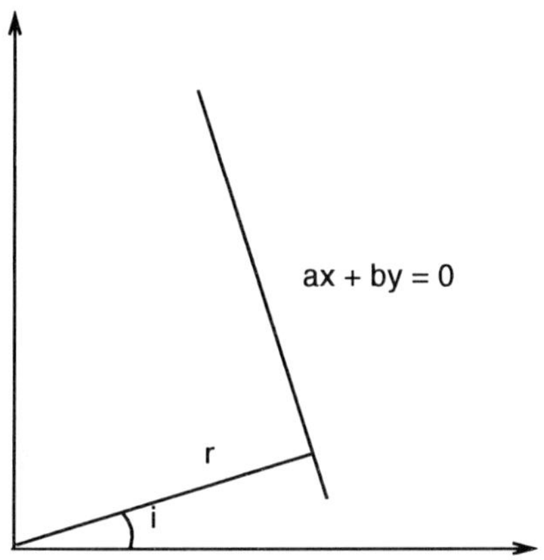

Fig. 4. The interface of an object can support multiple views of the object's state. For example, a line can be represented by its slant and its distance from the origin, or by the coefficients of its equation. Although a line object would use only one of these representations internally, it could support both views externally.

The concepts of object *type* and *class* are closely related to data encapsulation. While the class of an object in a class-based system usually describes every aspect of the object, including its internal data structures and algorithms, the type of an object describes only the external interface, i.e. the object's behavior. Hence, the class is associated with *implementation* issues, while the type is associated with object *specification.*

A *subtype* is a special case or refinement of its base type. The power of subtyping stems from the fact that the external interface of a subtype object is upward compatible with the interface of the supertype objects. In particular, a subtype object has (at least) all the methods of the base type, although the actual operations performed by these methods may have been redefined for the subtype objects. Such redefinable methods are often called *virtual functions.*

The image-with-histogram discussed in the previous section would typically be a subtype (and probably also a subclass) of an ordinary 2-D image. That would allow all existing filters to be applied to this new image, without any modification to the source code. The pixels would obviously have to be accessed through virtual functions, which can be redefined to keep the histogram up-to-date.

In general, the use of virtual functions implies *late binding*: when a method is applied to an object, the system has to examine the type of the object at run-time in order to determine which actual function to invoke. In strongly-typed languages, the compiler is sometimes able to deduce the exact object type, and can generate code for calling the correct actual function directly. In that case, there will be no run-time overhead. On the other hand, such code can never handle new object types without recompilation. Thus, there is a trade-off between run-time overhead and system flexibility.

Analogous to subtypes, a *subclass* is a specialization of its base class. However, the main benefit of subclassing is that it allows code to be reused through an *inheritance* mechanism.[6] If new objects for network communication can be implemented as subclasses of existing classes, a degree of code reuse can be achieved also at the communication level, and not only at the level of complete modules.

In a sense, the hierarchy of object types gives rise to a hierarchy of protocols: if objects of some specific type T are transferred under a certain protocol P, then clearly it should possible to use objects of any subtype $T_1 \subseteq T$ under P. For modules *requiring* the specialized objects T_1, it is natural to define a *subprotocol* $P_1 \subseteq P$, which only allows objects of type T_1 or more specific types. The subtyping rules for functions then also apply to modules.[6] For example, any module M with input protocol P_{in} and output protocol P_{out} could be replaced by a module M' with input protocol $P'_{in} \supseteq P_{in}$ and output protocol $P'_{out} \subseteq P_{out}$.

However, all this requires that objects can be transferred between modules with their original types retained, which presents an interesting problem since modules in most network systems execute in separate address spaces. Some possible solutions will be discussed in the next section.

4. Persistent Objects in Distributed Systems

In general, objects are destroyed when the process in which they were created is terminated. Objects are also bound to their original address space since they are often referenced by other objects through pointers. Furthermore, their internal data structures usually contain in-store pointers, e.g. to strings and other objects. *Persistent* objects, on the other hand, are more or less independent of processes and address spaces. They can live through several runs of the same program and be moved between programs executing in separate processes. Therefore, persistent objects is a good basis for inter-process communication.

There are several problems with implementing persistent objects that have to be solved. As mentioned before, pointers containing object addresses are valid only in a particular address space and must be replaced by some sort of object identifiers when objects are moved from one process to another. This immediately raises the question of how to identify objects globally, and how an object can be found from its identifier. Should the methods be stored with the object, and if so, how should they be represented? How can classes and class hierarchies be represented externally? Should it be possible to interact with an externally stored object, or should objects always be fetched into the address space of a process before they are accessed? If an object which contains references to a large number of other objects is recreated in a new address space, should all referenced objects also be recreated at the same time, or not until they are accessed?

A number of languages and systems which support persistent objects have been developed over the last few years. Some of them can be thought of as extensions of a programming language, for example Smalltalk-80,[7] Objective-C[8] and Eiffel.[9] In all these language extensions, the objects are "linearized" and stored as byte arrays. The external description typically includes the object's type and internal data structures, while its methods are only implicitly represented by the type tag. The objects are thus "frozen" externally, and must be fetched into the workspace memory of another process before they can respond to method calls. In all these cases the external representation of objects is closely related to the object implementation in a particular programming language.

Other solutions come from the field of object-oriented databases. Such systems usually require a separate database server and are often transaction-oriented. The objects typically exist only in the database and information about them is obtained through queries sent over the network. In some distributed object-oriented databases, mechanisms for actually *transporting* objects across networks can be found. One example is the Object Exchange Service used in the Zeitgeist database system.[10]

For the purpose of object-oriented communication in computational networks, we will stick to the simple stream-based approach since computational network systems rarely use centralized databases as the primary information exchange mechanism, mainly because of the limited bandwidth. The next section describes an implementation of persistent objects which operates close to the programming language level, yet is not committed to a particular language or a particular object implementation.

5. XOR

XOR (eXternal Object Representation) is a simple scheme for implementing persistent objects. It allows programmers to transfer conveniently a single object or a cluster of related objects from one address space to another. Processes may communicate through a variety of mechanisms, such as files, pipes, shared buffers and clipboards (see Fig. 5). XOR encapsulates these mechanisms in subclasses of an abstract stream class, which allows new transport layers to be used with existing application code.

Every object handled by XOR is required to provide a description of its internal data structures. These descriptions, which are usually implemented as function call sequences, will be used by the I/O system to read and write objects. The names and arguments of such *data description functions* are programming language dependent and XOR provides one library for each supported language.

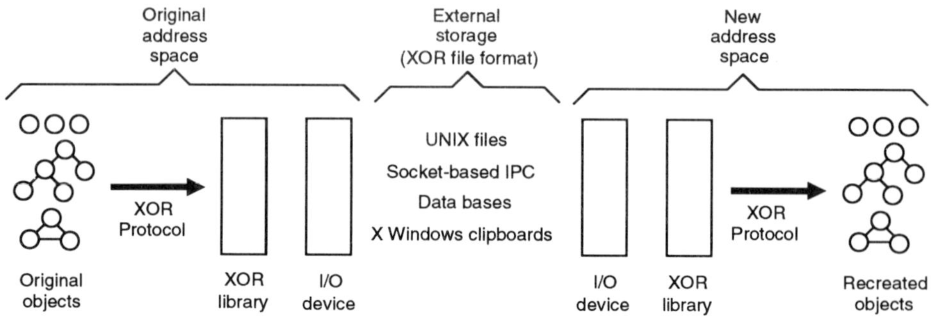

Fig. 5. XOR allows objects to be transferred between address spaces by a variety of communication mechanisms.

5.1. The XOR Protocol

The basic idea in XOR is to describe the information contained in the objects using a simple protocol. XOR goes beyond conventional approaches to object persistence because it is intended to make object descriptions language independent, thereby allowing objects to be moved freely between programs implemented in different languages. The XOR protocol also allows data to be moved between architectures with different representation of integers and floating-point numbers. This is accomplished through a separation of the XOR *protocol*, which defines *what* can be stored, from the physical *file format*, which determines *how* data is actually encoded.

XOR can be viewed as an extension of XDR (eXternal Data Representation) which is a DARPA standard for data encoding.[11] The XDR protocol is defined as a data declaration language, similar in flavor to C. The primitives of this language are simple data types such as integers, floating-point numbers and strings. With those primitives and some compound types, more complex data structures can be described. For example, a simple image could be described in XDR with:

```
struct {
    unsigned int xsize;
    unsigned int ysize;
    bytes pixels<1024*1024>;
}
```

where *pixels* is declared to be a pixel buffer with a maximum size of 1 Mbyte[a]. This abstract data declaration will eventually be mapped to function calls implementing the data encoding/decoding. For the example above, the calls to an XDR support library for C could look like this:

[a] This maximum size will be part of the protocol and used for error checking, but can otherwise be chosen arbitrarily.

```
xdr_u_int(stream, &pic.xsize);
xdr_u_int(stream, &pic.ysize);
xdr_bytes(stream, &pic.pixels, &pic.phys_size, 1024*1024);
```

where `pic` is a C structure and `stream` is an XDR data stream descriptor. These lines of code will be executed whenever `pic` has to be written to or read from the stream. The actual XDR function calls may look very different in another programming language—only the XDR data *declaration* is language independent.

Note that the same function calls can be used for both encoding and decoding. The actual operation performed will depend only on the state of the I/O system. This makes it easy to keep reading and writing consistent.

XOR extends the XDR protocol with class references and references between objects. However, XOR files contains only XDR formatted data.

5.2. Class Representation

The first and most important problem with persistent objects is to represent their classes (or types) externally in an unambiguous and space-efficient way, using some sort of class identifier. A program reading an XOR file has to map each class identifier to a suitable constructor in order to recreate the objects with their original class retained (i.e. the class they belonged to in the original address space).

In XOR, classes are encoded as variable-length strings, typically containing the print-name of the class. Various compression techniques used in XOR significantly reduce the space overhead caused by these tags. This will be discussed further in Section 5.4.

The mapping of class tags to object constructors is implemented with a dynamic look-up table, which can be extended at run-time. No source code has to be changed when derived classes are created, and existing programs can handle files containing objects of newly defined subclasses without recompilation. Of course, the programs have to be linked with the code implementing the new classes, but this can be done at run-time by a dynamic linker, e.g. GNU's DLD[12] or the `dlopen` facility in UNIX System V (see e.g. the Solaris operating system[13]).

Together with the class tag, two version numbers (*major*, *minor*) are stored. The XOR library compares them with the version of the code implementing the class and can reject obsolete data files.

5.3. Object References

Objects often refer to each other by addresses. Such pointers must be replaced when the objects are encoded, since the old addresses will not be valid in the new address space. The resulting graph of objects and inter-object references may contain cycles (see Fig. 6). The receiver process must not change that graph structure.

For this reason, XOR creates a one-dimensional array of objects and replaces all object pointers with indices into this array. It is then vital that all objects in the object array originate from the same memory address space, since the index representation

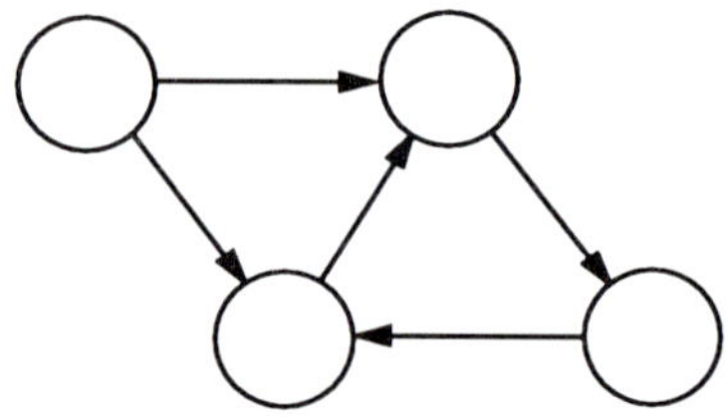

Fig. 6. Pointers between objects can form cyclic references. Such cycles must be saved and restored properly by a persistent object mechanism.

would otherwise be ambiguous. Therefore, XOR exchanges *packets* of objects with the client program, and the object identifiers are local to each packet. Hence, objects referring to each other by addresses must be sent in the same packet. Actually, all objects referenced by some object in a packet will automatically be pulled into that packet during encoding, which means that the whole *object closure*[b] is saved. A stored packet cannot be altered in any way, unless it is read back into workspace memory so that all its internal pointers can be resolved.

The XOR declaration of a pointer-to-object variable includes a *reference count*, which is the number of times the referenced object will be deleted through the pointer being declared. Usually, this value is either one (if the owner object is responsible for destroying the referenced object) or zero. These numbers are used by the XOR system to compute the total number of times each decoded object is expected to be deleted, and set the reference counter in the decoded object accordingly[c].

5.4. Space Overhead

The space needed for representing object types and references externally can be significant. Unless great care is taken, the extra cost in bandwidth will be prohibitive, especially for the kind of massive data structures (10 000 objects or more) that are common in computer vision applications.

XOR files uses the XDR encoding, which is a binary file format and therefore reasonably compact. Since the objects in a packet are always encoded and decoded together, the packet is a natural candidate for compression. XOR uses two simple compression techniques, which take the typical structure of XOR files into account and which seem to be very effective in practice: *class tag compression* and *redundant pointer compression.*

---

[b] The object closure is the set of objects that can be reached from an object through pointers, either directly or indirectly via other objects.

[c] In the C++ binding of XOR, the `Xor` base class has no reference counter. Therefore, the total number of expected `delete` calls for an object must equal one.

Although the total number of existing classes is large, the number of different classes represented in a single packet is usually very small. Therefore, strings with typically 10-30 characters identify the classes globally, but they are mapped to integer tags that are local to the packet. Hence, the full string tag does not have to be stored more than once in each packet. Furthermore, two or more consecutive instances of the same class can share a single (integer) class tag. In most cases, this makes the space required for class tags small. For example, a computer vision application may need to store thousands of contour patches on file. If they all belong to the same class, the space required for the single necessary class tag and the time spent searching for the corresponding constructor when the objects are decoded will be negligible.

In general, pointers between objects must be represented by one integer each. However, such pointers are often redundant. Consider for example the polygon object implemented as an array of pointers to vertex objects in Fig. 7a. Since the polygon could also be represented by an array of vertices (i.e. the vertex object addresses could be *computed* directly from the index) the pointers carry no structural information in this case and the representation in Fig. 7b is therefore equivalent. This redundancy

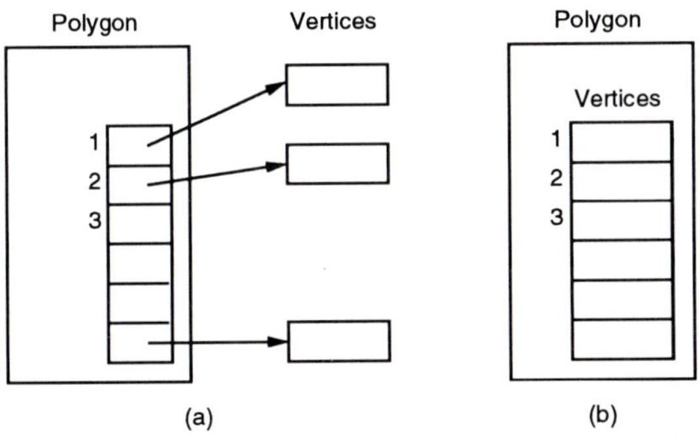

Fig. 7. Example of redundant pointers. In (a) the pointers carry no structural information, since the representation in (b) is equivalent.

shows up as a regular pattern in the list of object identifiers (indices) that XOR substitutes for the pointers, and the list is therefore stored run-length encoded. This simple trick significantly reduces the space overhead in common cases, but it still allows pointers to be completely independent in other situations. It also seems to do a fairly good job with mixed cases, where pointers are partially redundant (regular) and partially independent (random).

The number of bytes actually stored for each XOR/XDR type is fixed and does not depend on the architecture of the machine. XDR uses four-byte blocks, which means that all XDR types are represented by a multiple of four bytes on file. This can sometimes be a waste of space since e.g. single characters (not strings) will occupy four bytes each. However, this seems to be a minor problem in practice since most data types (long integers and floating-point numbers) are at least four bytes long. The four byte block size was chosen as a compromise between the space requirements and the run-time overhead caused by alignment problems.

5.5. *A C++ Interface*

In the C++ binding of the XOR protocol, all persistent objects must belong to a class derived from an abstract `Xor` class. The objects' internal structure is described by *data description function* calls from a special XOR constructor. For example, consider an abstract `Picture` class, with two derived classes: `Pixmap`, which contains the actual pixels, and `Window`, which is a subwindow of another `Picture`. Fig. 8

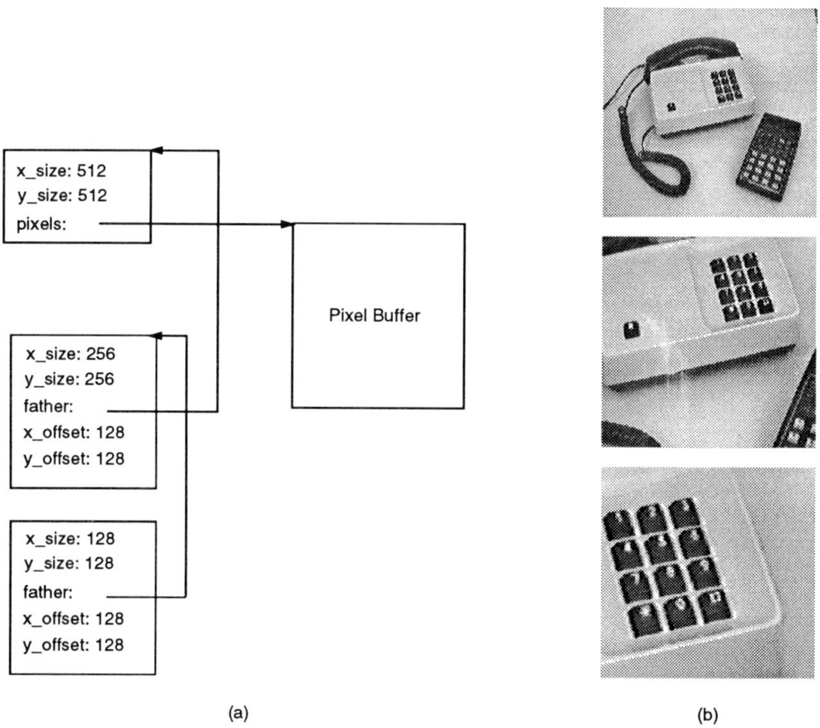

Fig. 8. (a) shows a small instance hierarchy consisting of one physical image object (of class `Pixmap`) and two logical windows (of class `Window`), which represents the images in (b). The `Pixmap` and `Window` classes are both derived from a common `Picture` base class.

shows a small object hierarchy, where two `Window` objects have been created from a

Pixmap. One possible C++ declaration of these classes is shown below.

```
class Picture : public Xor
{
    XOR_DECLARE(Picture, Xor);      // Declarations needed by XOR
    unsigned int x_size, y_size;
protected:
    Picture(XorPacket& packet) : Xor(packet) // XOR constructor
    {
        packet ^ x_size ^ y_size; // Save or restore size
    }
};

class Pixmap: public Picture
{
    XOR_DECLARE(Pixmap, Picture);
    char *pixels;
protected:
    Pixmap(XorPacket& packet) : Picture(packet)
    {
        unsigned int size = x_size * y_size;
        packet ^ xor_buf(pixels, size, x_size * y_size);
    }
};

class Window : public Picture
{
    XOR_DECLARE(Window, Picture);
    Picture *father;
    unsigned int x_offset, y_offset;
protected:
    Window(XorPacket& packet) : Picture(packet)
    {
        packet ^ father ^ x_offset ^ y_offset;
    }
};
```

The XOR constructors are used for both object encoding and object decoding. As with all C++ constructors, they are automatically called in superclass-to-subclass order, so each constructor only has to deal with the data fields of the class it belongs to.[14]

The XOR operator (^) does the actual I/O. The first operand is always an XorPacket, while the second operand can be any integer, floating-point, string or object pointer variable. The ^ operator is left-associative and returns the first operand (the packet), which makes it possible to have several operator calls in the same statement. Since the ^ operator is binary, special functions must be used when multiple arguments are needed to describe a data member. For example, the size of the pixel buffer in the `Pixmap` class above must be specified together with the `pixels` pointer in an `xor_buf` call:

```
packet ^ xor_buf(pixels, size, max_size);
```

The third argument `max_size` is a maximum size which is only used for error checking.

When a window is decoded, its `father` pointer will be resolved when the referenced `Picture` object has been recreated. Note that if a `Window` object is transferred with XOR, the whole chain of subwindows up to the `Pixmap` root will be restored in the new address space. For simplicity, a few details have been left out in the sample code. In particular, we would need virtual functions in the `Picture` base class for accessing the pixels and for creating new windows.

Apart from the `Xor` base class, the XOR/C++ library also contains a number of `XorStream` subclasses, which represent the various types of transfer mechanisms available, e.g. `XorInputFile` and `XorOutputFile` for ordinary files.

At the receiver end, the decoded objects are always stored on the heap, since their sizes are not known in advance[d]. In general, input objects must be down-cast by the decoding process, which involves run-time type checks. As an example, consider the following filter program which defines subwindows for all pictures in an XOR stream:

```
XorInputFile infile(infile_name);    // Open input file
XorOuputFile outfile(outfile_name); // Open output file

while (!infile.eof()) // While there are more objects in infile
{
    Picture *pic;
    infile >> pic;

    Window *win = pic->new_window(x, y, xsize, ysize);
    outfile << win;
    delete win;
}
```

While more data remain on the input stream, a new object is read. The `>>` operator verifies that the object actually is a `Picture` object (or a subtype thereof). A logical window is created from the picture and encoded by the `<<` operator.

5.6. A C Interface

There is also a C interface to XOR, based on the OPTIC (Object Programming Tools In C) package.[15] OPTIC allows classes and methods to be defined and used in C programs. It supports (single) inheritance, exception handling and polymorphism.

---

[d] Objects belonging to new subclasses usually contain new data fields and are therefore larger that the objects of the expected class.

The OPTIC macros require an ANSI C compiler, but no special preprocessing; OPTIC programs are written in pure ANSI C.

There are OPTIC classes corresponding to the `Xor`, `XorPacket` and `XorStream` classes found in the C++ interface. For historical reasons, they are called `Ext`, `Packet` and `PacketStream`, respectively. As an example, consider a polygon implemented as an array of points. The points can be represented by an `Ext` subclass, called `Point`, with the following OPTIC object declaration:

```
#define Point_Is Derived_From (Ext)

DeclareClass (Point,
{
    Constructor (New, (double x, double y));
},
{
    double x;
    double y;
});
```

where `x` and `y` are the coordinates of the point and `New` is a constructor. The following macro call defines the XOR representation of the `Point` objects:

```
DefineXor (packet,
{
    xor_double(packet, &self->Point.x);
    xor_double(packet, &self->Point.y);
});
```

The `xor_double` function corresponds to the overloaded `operator^(double)` in the C++ library. For the polygon class, which could be declared with:

```
#define Polygon_Is Derived_From (Ext)

DeclareClass (Polygon,
{
    Constructor (New, (unsigned int sz));
},
{
    Point *points;
    unsigned int size;
});
```

a single function call would do:

```
DefineXor (packet,
{
    xor_members(packet, &self->Polygon.points, &self->Polygon.size);
});
```

where `xor_members` registers an array of pointers to member objects.

5.7. *Summary*

To summarize, the advantages gained with XOR are:

- Network transparency: XOR removes the barrier between workspace memory and external storage. Complex structures of objects can easily be moved between processes.
- Incremental I/O design: With XOR, the I/O can be implemented gradually, as more classes are defined. New subclasses only need to handle the fields they define themselves; fields they have inherited are transferred by the XOR data description functions belonging to their superclasses.
- Distributed I/O design: No centralized code for reading and writing large object structures is needed. Instead, the I/O code is distributed over all objects involved, which significantly increases the modularity of the I/O system.
- Code reuse: Once the external representation of an object has been specified with the XOR protocol, the description can be used for all types of transmission mechanisms. Thus, if objects can be read from and written to files, they can also be transferred in shared memory buffers, X Windows[e] clipboards, and so on.
- Flexibility: An XOR library can handle instances of new subclasses without recompilation, and the code implementing new classes can be loaded at runtime. The XOR protocol also supports version control, so that object definitions can be changed, while still allowing old data files to be processed.
- Language independence: The XOR definition of an object does not depend on any particular programming language. Migration to new languages becomes easier with XOR, since programs written in different languages can cooperate.
- Efficiency: The objects read by XOR will be recreated using the most suitable object mechanisms available in the target language. For example, the C++ binding of XOR uses C++'s built-in classes, while the C library uses OPTIC.

6. Using XOR in Computational Network Systems

This section will demonstrate how XOR can be integrated into existing computational network systems, such as MNT, AVS, VIPWOB and KHOROS. A full description of these systems can be found elsewhere in this book. Here, we will concentrate on the communication mechanism. XOR is an integral part of MNT, and this system will be discussed first. XOR has also been used with AVS, which has a rather different stream model. Finally, we will indicate how XOR could be used with VIPWOB and KHOROS.

[e] X Windows is a trademark of M.I.T.

6.1. MNT

An MNT[4] module is implemented as a C++ subclass of the abstract `MntModule` base class, and is compiled with a `main` into an executable program. An MNT network consists of a set of processes, each executing such a module program.

MNT modules communicate through *ports*, which are represented by C++ port objects. There are different port subclasses for different communication protocols, all derived from the abstract `MntPort` base class. One of these subclasses is `MntXorPort`, which implements the XOR protocol. This class provides functions for sending and receiving XOR objects. Naturally, MNT uses the C++ interface to XOR.

For example, the `Picture` object described in Section 5.5 could be sent to an output port and read from an input port with:

```
void SendPicture(MntXorPort port, Picture *pic)
{
    port.Write(pic);
}

Picture *ReadPicture(MntXorPort port)
{
    Xor *object = port.Read();
    return Picture::cast(object);
}
```

where the `Picture::cast` function verifies that `object` really is a `Picture` or belongs to a `Picture` subclass.

All communication between modules and the MNT supervisor process and between most user-defined modules is object-oriented. MNT modules consume objects from their input ports, and produce new objects on their output ports. Objects in the communication channels are persistent and are therefore independent of the modules.

6.2. AVS

AVS, which is a commercial visualization system from Advanced Visual Systems Inc.,[2] takes a slightly different approach to module communication. Just as in MNT, modules usually run in separate processes, although multiple modules can share a single process in some cases. Often, a module is executed synchronously like a subroutine: it is invoked whenever the value on one of its input ports has changed, to compute a new output value. Therefore, data structures can be stored in buffers associated with the input and output ports. Alternatively, a module can execute asynchronously, which allows it to produce output data at any time, not just when new input data has arrived. However, it will then be the responsibility of the module to synchronize its operations with the rest of the network through calls to the AVS *flow executive*, which supervises network execution. Failure to do so may cause a loss of data if the module produces output data faster than the receiving modules can consume it.

AVS has a set of built-in, fairly general data types. There are *scalar values* (integers and reals), *strings* and *fields.* A field is basically a multi-dimensional array of scalars or vectors. It may also include a mapping from the "real-world" coordinate system to the array indices, and such mappings can be uniform or irregular. There are also more specialized data structures, such as *colormaps* and *geometry lists*, which are used for rendering, and *molecule data* for applications in the field of chemistry. Each type has its own representation in the C program.

However, there is currently very little support for user-defined data structures (AVS version 4). The user may define a `struct` in a special header file, which can be parsed (in run-time) by AVS. However, only a very small subset of the C syntax may be used in the declaration. In particular, pointers, enumerations and preprocessor symbols are not allowed.

It is a simple matter to define a new XOR `PacketStream` subclass, on top of a one-dimensional byte field. Actually, there is a predefined packet stream in OPTIC called `PacketMemory`, which reads and writes from memory buffers. From this packet stream two new subclasses, here called `AvsPacketInputField` and `AvsPacketOutputField`, can easily be derived and used for reading and writing, respectively. Given these subclasses, an AVS subroutine module which calls the virtual `print` function of incoming XOR objects could be coded in C as follows[f]:

```
int print_xor(AVSfield_char *input_field, int verbose)
{
    AvsPacketInputField ibuf;
    InputPacket ip;
    unsigned int i;

    ibuf = New (AvsPacketInputField, (input_field));
    while (!packet_stream_eof (ibuf))       /* until end of buffer */
    {
        ip = New (InputPacket, (ibuf));     /* next input packet  */
        for (i=0; i<packet_size (ip); i++) /* get packet contents */
            optic_print (packet_get (ip, i), stderr, verbose, 0);
        Delete (ip);
    }

    Delete (ibuf);
    return 1;
}
```

Output streams are handled similarly.

`AvsPacketInputField` and `AvsPacketOutputField` use variable-length byte fields. However, it may be more efficient to use a fixed-size AVS buffer for transferring XOR data, especially if the buffer can be shared between processes. Since the packet size

[f] This is the *compute* function of an AVS subroutine module.

Fig. 9. XOR packets can be broken into fixed-size chunks that fit into a shared memory buffer by a specialized XDR/XOR stream.

is not known in advance, and we do not want to fixate an arbitrary maximum buffer size at compile-time, the modules must in this case be executed asynchronously so that the packet stream can divide the XOR data into a number of buffer transfers (see Fig. 9). A special XOR stream, which uses the AVS functions `AVScorout_wait` and `AVScorout_mark_changed`, must be defined for this purpose.

6.3. VIPWOB and KHOROS

VIPWOB (VIsion Programmers WOrkBench) is another computational network tool, developed at the Laboratory of Image Analysis, University of Aalborg, Denmark.[3] The library provides special versions of the UNIX system calls `open`, `close`, `read` and `write` that are network transparent and can be used for all module-to-module communication. Since there are predefined XOR streams based on these system calls, it would be very easy to adopt XOR as a high-level communication protocol in VIPWOB applications.

The same applies to the KHOROS system,[1] since its communication is also based on standard I/O mechanisms and temporary files.

7. The Flavor of Object-Oriented Communication

In this section we will take a look at some of the advantages gained by using persistent objects as a basis for communication in computational networks. Several software techniques such as dynamic linking and daemons are very straight-forward to use in an object-oriented framework.

7.1. User-Definable Types

Undoubtedly, one of the most important feature of the object-oriented communication scheme described above is that it allows user-defined data types to be transferred over the network in a flexible yet principled way. By carefully designing the external interface of an object, the programmer can impose certain constraints on the object's internal data structures no matter where in the network the object is processed. Moreover, the general subtyping scheme discussed in Section 3 allows existing modules to be used with new, specialized objects types. This reuse of code will not only cut development time for new system prototypes, but will also make the module pool more manageable by significantly reducing the total number of modules required. The two most common situations where new subtypes are needed are:

- when more information has to be put into an object, and
- when the behavior of an object has to be changed.

For example, a user might want to extend the `Point` object from Section 5.6 with a positional uncertainty value (again using XOR/OPTIC):

```
#define MyPoint_Is Derived_From (Point)

DeclareClass (MyPoint,
{
    Constructor (New, (double x, double y, double variance));
},
{
    double v;
});
```

Assume that `Point` has been derived from a `Drawable` class (and not directly from `Ext` as in Section 5.6) in which a postscript[g] method has been declared. The user can then redefine it to draw a circle around the point, with a radius computed from the variance:

```
DefineMethod (void, Drawable, postscript, (FILE *file),
{
    CallSuper (Drawable, postscript, (file)); /* superclass version */

    fprintf (file, "newpath %f %f %f 0 360 arc stroke\n",
             self->Point.x, self->Point.y, sqrt(self->MyPoint.v));
});
```

With these definitions, the new point objects can be sent to an existing display module which calls the postscript methods of all drawable input objects. The circle around the points constructed above will the appear in the graphics window of the display module (see Fig. 10).

7.2. Monitoring Network Communication

Although XOR is a binary data format, the communication protocol is well-defined in the sense that objects can be decoded, processed and forwarded by any module in the network, no matter what classes the objects belong to. This makes it possible to listen to any communication channel in the network and find out what is going

[g] Postscript is a trademark of Adobe Systems Inc.

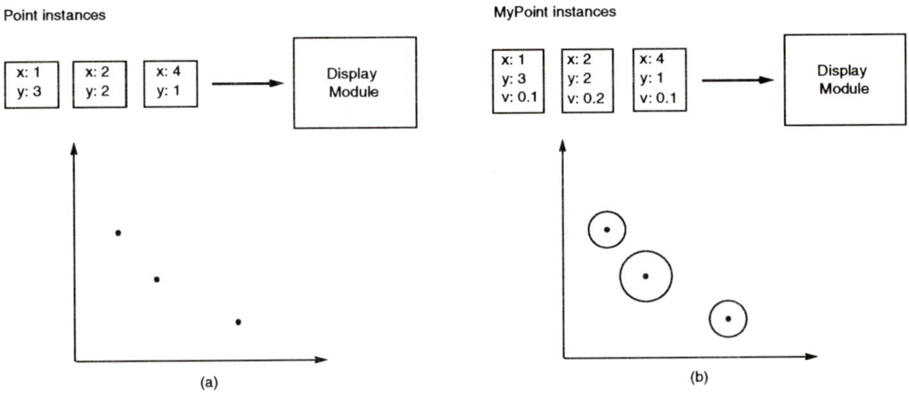

Fig. 10. (a) A display module which calls the `postscript` method of the objects it receives, and the resulting output graph. (b) When objects of a new subclass are sent to the same module, they are displayed differently.

on. For example, a log module can intercept all objects sent in the channel, call their print method[h], and display the output in a window. Fig. 11 shows an example of this, where the communication between an MNT application module and a parameter repository module is logged.

7.3. Merging Data Streams

Data streams in a computational network often have to be merged into a single stream. This can be done either by special modules, as in AVS, or transparently by the network system, as in MNT. In either case, merge operations must be done according to a protocol, and not just blindly at the byte-level. For example, when merging text streams, one usually avoids splitting input lines. With an object-based communication protocol it is natural to merge data streams at the level of objects, and thereby ensuring the integrity of the objects.

7.4. Object Closure

Since pointers between objects are automatically resolved by XOR, it is very easy to send clusters of related objects between modules. For example, the relations between an image, polygon contours extracted from it, and straight-line approximations of the polygons could be represented by pointers, as in Fig. 12. The entire object graph can be transferred in one XOR function call.

[h] The C++ `Xor` class declares a virtual print function, and all OPTIC objects have a print method declared in the root class.

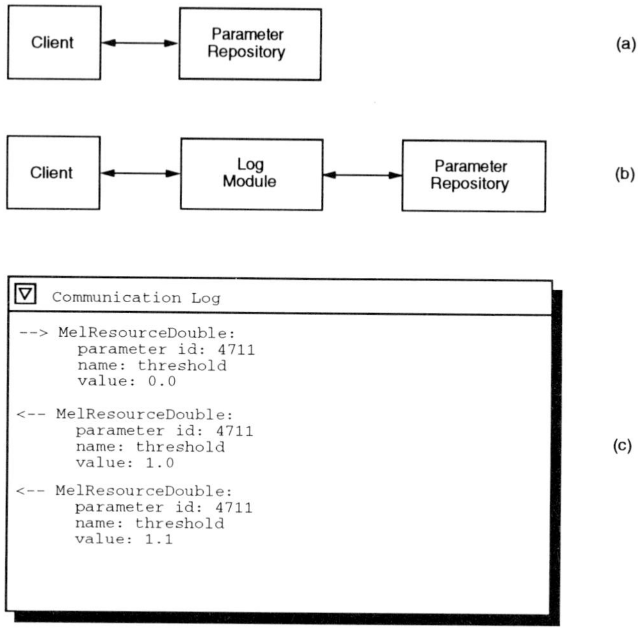

Fig. 11. (a) shows a client module connected to an MNT parameter repository module. The client sends parameter definitions to the repository, which displays the parameters as glyphs in a window on the screen (e.g. as scrollbars). Each time the user changes a parameter value, a message is sent back to the client module. This communication consists of objects (of class **MelResourceDouble** in this case) which can be intercepted by a log module, as shown in (b). The log module calls the print method of the objects and displays the output in a separate window (c).

7.5. Dynamic Loading

As mentioned in Section 5.2, XOR uses a dynamic look-up table to map class tags to object constructors. There is one entry in this table for every class known to the XOR library, and more classes can be entered at any time.

However, an executable program cannot reach the methods of newly defined object classes unless it has been *linked* with the compiled code which implements the classes. Either the executable program has to be completely relinked and restarted, or the new code segments have to be linked in at run-time. In the latter case, the name of a variable or function through which all information about a new class can be reached may be derived directly from the class name stored with the object. For example, if a program using XOR/C++ does not find a match for a class name

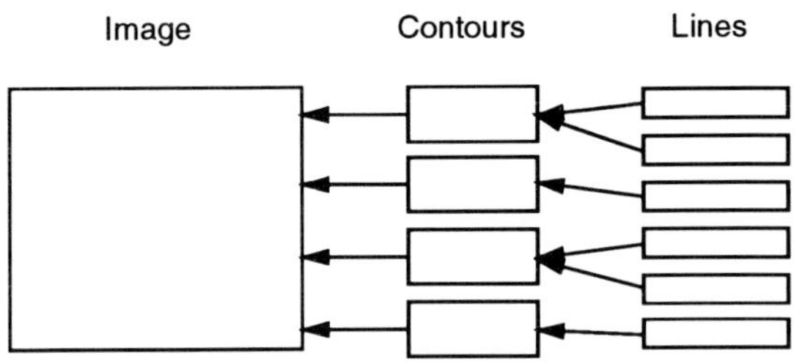

Fig. 12. The relation between an image, the contours extracted from it, and the corresponding (piece-wise) straight-line approximations may be represented by pointers. The resulting object cluster can be transferred between the modules as one entity .

"Foo" in the global table of XOR constructors, it will try to dynamically link in a function `Foo::xor_init`. If that function is called from the program, it will (among other things) add the missing "Foo" → `Foo::Foo(XorPacket&)` mapping to XOR's look-up table.

The executables still have to know which library files to search for missing classes. In an implementation of dynamic linking for stand-alone programs based on the `dlopen(3)` dynamic linker interface of System V UNIX, the library file names were specified by the user with command line options:

```
cat file.xor | drawable-to-postscript -dll my-data-structures | lpr
```

Here it is assumed that `my-data-structures` is the name of a user-defined library of class definitions. In a computational network like MNT, these file names could be specified as *module resources* instead of command line options.[4] Alternatively, a module could broadcast a request for the name of the implementation file when it receives an object it does not recognize. Provided that the object was originally created within the network, some other module must know which class implementation file to load.

7.6. Procedures vs. Data Structures

The combination of persistent objects and dynamic (run-time) linking blurs the distinction between procedures and data. For example, suppose that the internal state of a module has to be monitored. A common way of doing that is to install a *daemon* into the module. The daemon will wait for a specific event to occur, and will react by sending a message to its creator. With persistent objects available, a daemon object can be created at run-time, just like any other object, and be sent to the destination module through a channel, provided that an XOR description exists for the daemon class. The daemons typically enter the destination module through a special port, and are put on a waiting list. The module can then periodically call a daemon method, `Execute`, to allow each daemon to check for the events it is interested in (see Fig. 13).

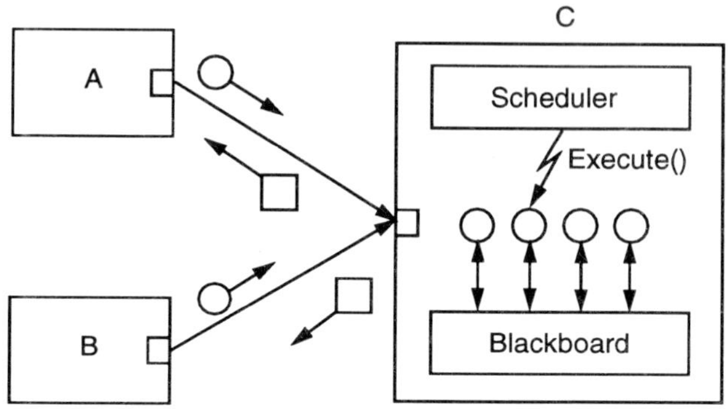

Fig. 13. Modules A and B send daemon objects (depicted as circles) to module C. Each daemon waits for a particular condition to be fulfilled by C's internal data structures and reacts by sending messages (depicted as squares) back to its creator.

8. Source Code Availability

The XOR software is distributed freely under the terms of the GNU Library General Public License as published by the Free Software Foundation. It is part of the CVAP software distribution, which is available by anonymous ftp from ftp@bion.kth.se.

The current CVAP distribution (version 2.0) contains an XOR/C++ library that can be built with the GNU g++ compiler, version 1.40. However, the C++ code in this paper is compatible with a new XOR/C++ library that relies on parameterized types and which can be built with cfront 3.0.2 or later. This version will soon be available in CVAP distribution 2.1 at the same ftp site.

The C binding of XOR is part of the OPTIC package, which can also be found in the CVAP distribution. A technical report describing the XOR protocol and file format can be found on directory cvap/2.0/optic/doc/xor-report.

The software currently runs under SunOS 4.1 and 5.2. However, it has been designed to be portable, so it should run under any System V Release 4 UNIX. Please direct all inquiries about the software to candela@bion.kth.se.

9. Summary

The most important concepts from the field of object-oriented programming have been introduced. We have tried to show how computational networks could benefit from an object-oriented communication mechanism, especially with respect to data encapsulation and subtyping. A scheme for implementing persistent objects in a language independent way has been presented, and we have indicated how this could be incorporated into existing computational network tools. Finally, some examples of object-oriented design in computational networks have been given.

Acknowledgements

The support from the Swedish National Board for Industrial and Technical Development (NUTEK) is gratefully acknowledged.

References

1. J. Rasure, D. Argiro, T. Sauer, and C. Williams. A visual language and software development environment for image processing. *International Journal of Imaging Systems and Technology*, 2:183–199, 1990.

2. Advanced Visual Systems Inc. *AVS Programmers Manual*, 1992.

3. N. O. Kirkeby and H. Christensen. Vipwob description and reference manual. Technical report, Laboratory of Image Analysis, Universisty of Aalborg, Denmark, April 1992.

4. L. Olsson. Module network tool. Technical report, Dept. of Numerical Analysis and Computing Science, Royal Institute of Technology, Stockholm, 1992.

5. M. Rendahl, J. Gårding, and T. Lindeberg. Candela—an overview. Technical Report CVAP-TN09, Dept. of Numerical Analysis and Computing Science, Royal Institute of Technology, Stockholm, oct 1990. (Presented at the Aalborg University Vision Workshop Nov 22, 1990).

6. G. Blair, J. Gallagher, D. Hutchison, and D. Shepherd. *Object-Oriented Languages, Systems and Applications.* Pitman, 1991.

7. A. Goldberg and D. Robson. *Smalltalk-80: The Language.* Addison-Wesley, 1989.

8. B. J. Cox and A. J. Novobilski. *Object-oriented Programming: An Evolutionary Approach.* Addison-Wesley, 2nd edition, 1991.

9. B. Meyer. *Object-Oriented Software Construction.* Prentice Hall, 1988.

10. G. J. Pathak, J. Joseph, and S. Ford. Object exchange service for an object-oriented database system. In *Proceedings of the 5th International Conference of Data Engineering*, 1989.

11. Sun Microsystems Inc. *External Data Representation: Protocol Specification*, 1990. Also available at the ARPA Network Information Center as RFC1014.

12. W. W. Ho. An approach to genuine dynamic linking. *Software—Practice and Experience*, 21(4):375–390, 1991.

13. Sun Microsystems Inc. *SunOS 5.1 Linker and Libraries Manual*, 1992.

14. M.A. Ellis and B. Stroustrup. *The Annotated C++ Reference Manual.* Addison-Wesley, 1990.

15. H. Winroth. Optic reference manual. Technical Report CVAP95, TRITA-NA-P9127, Dept. of Numerical Analysis and Computing Science, Royal Institute of Technology, Stockholm, November 1990.

VISTA: A SOFTWARE ENVIRONMENT FOR COMPUTER VISION RESEARCH

ARTHUR R. POPE DAVID G. LOWE
Department of Computer Science, University of British Columbia
Vancouver, B.C., Canada V6T 1Z4
{pope,lowe}@cs.ubc.ca

ABSTRACT

Vista is a software environment supporting the modular implementation and execution of computer vision algorithms. Because it is extensible, portable, and freely available, Vista is an appropriate medium for the exchange of standard implementations of algorithms. This paper, an overview of Vista, describes its file format, its data abstraction, its conventions for UNIX filter programs and library routines, and its user interface toolkit. Unlike systems that are designed to principally support image processing, Vista provides for the easy creation and use of arbitrary data types, such as are needed for many areas of computer vision research.

1. Introduction

As a science concerned with creating artificial systems, computer vision ought to emphasize the evaluation and comparison of alternative methods[1,2]. It should be simple to compose large systems from components created by other researchers. That this is seldom done is partly due to the lack of a suitable, common software environment. Without a common environment, implementations and data can be exchanged only with great difficulty.

What is required of a software environment for computer vision research? It should support a large variety of data types, including not only images of various kinds but also other objects such as edge vectors, object models, and camera calibration parameters. It should be readily extensible in terms of both the data structures supported and the algorithms implemented. So that it can be understood easily, it should be modular, lightweight, and straightforward. It should have reasonable efficiency. Finally, to encourage its widespread adoption, the environment should be based on a popular language, be written for portability, and be made freely available.

Of these requirements, extensibility deserves special mention because it is difficult to provide and often overlooked. No design can anticipate the tremendous variety of data structures that researchers may wish to create; the design must, instead, provide a mechanism by which new structures can be defined, and then stored and manipulated alongside standard structures like images. It must be possible to define new structures out of existing ones (e.g., building an image pyramid from images) and to augment existing structures with additional information (e.g., annotating an image with information about how it was acquired). Moreover, extensions like these should not require modifications to existing software.

Several existing software environments provide some support for computer vision research, yet fail to meet all of the requirements stated above. Often these environments focus too narrowly on image processing (e.g., Khoros[3] and LaboImage[4]); while providing excellent support for image manipulation, they make few provisions for storing and manipulating other kinds of information (e.g., edge vectors). Some environments must be purchased, which limits the ability to distribute code developed within them (e.g., HIPS[5], HVision[6], and KBVision[7]), or they bear copyrights restricting how code may be copied, modified, and redistributed, which limits their extensibility (e.g., Khoros[3]). One promising system, the ARPA Image Understanding Environment[8], remains under development.

Faced with these shortcomings, we began three years ago to develop a new environment whose particular goals were to support computer vision research and to foster the exchange of research products. That environment, Vista, is the subject of this paper. Vista is now freely distributed with extensive documentation, a collection of image and edge vector manipulation routines, tools for printing and viewing images and edge vectors, and a toolkit to aid in developing interactive applications. As a foundation for developing and exchanging algorithms, Vista is designed to be simple, lightweight, and accessible; it requires no special hardware, nor is it intended for real-time applications. It is written in ANSI C, and it runs on most UNIX platforms and X Window System displays.

Unlike most environments for image processing, Vista emphasizes easy extension. It uses a flexible, self-describing file format so that objects of many types, including custom ones, can be represented for storage and interchange. In section 2 we describe the file format. Vista already includes support for some commonly needed types, such as one for representing images described in section 3. The environment can be extended readily to support new data types.

Vista operations are packaged both as stand-alone UNIX programs and as library routines. Section 4 illustrates how the programs can be used from a UNIX shell to manipulate files containing images and other objects. Section 5 describes the library, which can be used to construct custom applications. One component of the library is a toolkit for rapid development of X Window System applications. Using a few toolkit routines, and without knowledge of X programming, one can create interactive, menu-driven applications that display images and vectors. The toolkit is described further in section 6. An appendix summarizes operations implemented in the fall 1993 release of Vista.

2. File Format

Vista employs a data file format that is flexible, designed for extensibility, and largely self-describing. It is based on a simple structure called an *attribute list*—a sequence of attributes, each having a name and a value. An attribute's value may be a number, a string, or a keyword; it may be a nested attribute list; or it may be an object, such as an image, with an explicitly-named type and a value that, in turn, is specified by a nested attribute list. Attribute lists can thus represent a wide variety of data structures, as the example below will illustrate.

When an attribute list is stored in a data file, it is encoded primarily in ASCII. However, information that must be stored compactly (such as an array of image pixel values) is expressed using a machine-independent, binary encoding. Within a file, ASCII and binary portions are kept separate so that the ASCII portion can be examined directly by a human.

```
V-data 2 {
	ufo: image {
		data: offset of first image's pixel values
		length: length of first image's pixel values
		nbands: 1
		nrows: 256
		ncolumns: 256
		repn: float
		comment: "UFO sighted over Vancouver"
		pixel_aspect_ratio: 1.25
		icon: image {
			data: offset of second image's pixel values
			length: length of second image's pixel values
			nbands: 1
			nrows: 32
			ncolumns: 32
			repn: ubyte
		}
	}
	parameters: camera {
		data: offset of camera object's binary data
		length: length of camera object's binary data
		focal_length: 30.0
		center_x: 0.5
		center_y: 0.51
	}
}
^L
first image's pixel values
second image's pixel values
camera object's binary data
```

Figure 1. Example of a Vista data file. A list of attributes is encoded primarily in ASCII to allow human interpretation. Large volumes of data are encoded in binary, in a separate part of the file, for efficiency.

Figure 1 shows a data file that contains two images, plus a third object whose type is identified as `camera`. The first image, named `ufo`, contains several attributes describing it's size, pixel representation, and the location of its binary-encoded pixel data within the file. One of `ufo`'s attributes is actually another image, named `icon`. The `camera` object contains both numeric attributes and binary-encoded data. This example illustrates some key properties of Vista's data file format:

- Related objects can be grouped, either by nesting one object within another (as `icon` is nested within `ufo`) or by listing them sequentially within one file (as the `ufo` image is paired with the `camera` object).
- Because the format is self-describing, new types of objects can be introduced without changing or obsolescing existing software. Each object includes enough information about itself to permit it to be copied from one file to another unmodified, even by programs not equipped to interpret the object. In this example, the `camera` object is a custom one, newly created; however, it will be carried transparently by any existing programs invoked to transform the `ufo` image.
- Existing types can be extended as needed by inserting additional attributes. For example, if some application requires that images be annotated with some extra information, we can simply include that information in the form of additional attributes within `image` objects. The application can check for the new attributes while existing programs, not programmed to interpret the new attributes, will simply pass them transparently.
- Much of a data file's contents can be interpreted directly by a human. Using a text editor, one can edit attribute values, delete attributes, and insert new ones. (Offsets specifying the location of binary data remain valid despite insertions and deletions because they measure from the end of the ASCII portion of the file.)

3. Image Format

Images are one type of object for which Vista includes built-in support. The image type is flexible enough to represent many of the kinds of images most commonly encountered, yet it remains simple and efficient. A Vista image includes a three–dimensional array of pixels indexed by row, column, and "band" number. The third dimension, band, can be used to represent any of four "virtual" dimensions: the discrete frames of a motion sequence, various camera viewpoints, multiple color channels, or the real and imaginary components of complex pixel values. It can also be used to represent some combination of these four dimensions according to a fixed hierarchy — allowing, for example, a single image to represent a motion sequence of stereo pairs of RGB color images. Seven different pixel representations are supported, ranging from a single bit to a 32–bit signed integer and a 64–bit floating point number. Vista's image processing routines and programs manipulate multi-band images as readily as a single-band ones, and most support all pixel representations equally well.

A Vista image also includes an attribute list that may contain any number of attributes describing the image. Among these are standard attributes, such as `nrows` and `ncolumns`, specifying the dimensions of the image and indicating how its various bands should be interpreted. However, the list may include any other, custom attributes needed to support particular applications (e.g., for a photometric stereo application, images may include attributes specifying light source locations). Custom attributes are carried transparently by Vista software—whenever a new image is created by processing an old one, the new image automatically inherits the attributes of the old. Consequently, existing software does not generally need to be modified whenever custom attributes are employed.

Although Vista's image representation is unique, it is easily mapped to other formats. Vista includes software for translating images in both directions between Vista's format and Portable Pixmap (PPM) format. The PBMPLUS package[9] can be used to reach a large variety of other formats from PPM format.

4. Using Vista Programs

Most of Vista's data manipulation operations are available in the form of UNIX filter programs. Following the UNIX philosophy, each program is designed to perform just one particular task well. All programs follow a convention governing their command line arguments, allowing most options to be omitted or abbreviated, and all provide some on-line help. Some, like the viewer shown in figure 2, are interactive and have graphic user interfaces.

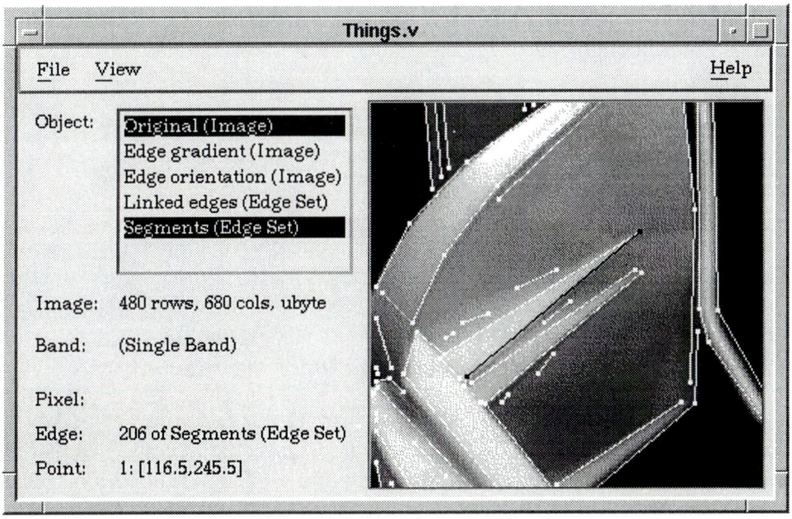

Figure 2. One of Vista's programs for viewing images and edge vectors.

A typical program reads objects from a file or stream, manipulates the objects somehow, and writes the result to another file or stream; thus, programs can be "piped" together to perform a sequence of operations. Here, for example, is a sequence that detects edge pixels, links neighboring edge pixels into chains, and breaks the chains into approximating straight line segments:

```
vcanny < image.v -sigma 1.0 -noise 0.5 |
   vlink -minlength 10 | vsegedges > segments.v
```

A program that processes one type of object will copy any other objects unchanged from input to output; thus, when files contain a mix of both standard objects (e.g., images) and custom ones (e.g., the `camera` object in figure 1), existing programs can still be used to process the standard objects. A program will also copy unchanged any unrecognized attributes, allowing even standard objects to carry customized information.

5. Programming with the Vista Library

Most of Vista's data manipulation operations are also available in the form of C-callable library routines that can be combined to build new modules. These routines, and the data structures they manipulate, are organized in an object-oriented fashion to hide nonpertinent implementation details. An image, for example, is represented in memory in a way that optimizes common operations; it includes an array of pixel values, various indexing arrays to speed pixel access, a structure with fixed fields for recording commonly-accessed image attributes, and a list containing any other image attributes. However, the programmer refers to this data structure by means of a single pointer and accesses it using a collection of macros and routines that shield most details. This is illustrated by the following code fragment, which zeros pixels in band 0 of an image:

```
VImage image;           /* handle to the image */
int row, column;        /* row and column indices */

for (row = 0; row < VImageNRows(image); row++)
   for (col = 0; col < VImageNColumns(image); col++)
      VPixel(image, 0, row, col, UByte) = 0;
```

A typical image processing routine accepts one or two source images and returns a destination image. By convention, the caller has a choice of either supplying the destination image, or letting the routine create and return a suitable destination image. This gives the programmer full control over how memory is allocated for images.

When an image is read from a data file like the one illustrated in figure 1, it is automatically converted to the more-efficient, internal representation; when written to a file, it is converted back again to the file format. Using facilities of the library, the programmer can define a new, custom type and register it so that objects of that type are automatically converted between the file format and an efficient, internal format. The library can thus be readily extended to recognize and support new types of objects. Among the library's other general facilities are routines for handling errors, parsing command line arguments, and generating PostScript output depicting images.

6. Creating Interactive Programs

When one is experimenting with an algorithm it is often helpful to be able to embed it within an interactive program that allows parameters to be adjusted and results viewed quickly. To aid this style of experimentation, the Vista library includes a package, called VX, for simplifying the task of creating programs that display images, have graphic user interfaces, and run under the X Window System. Using the VX package, and without knowledge of X Window System programming, one can create a program that will accept commands, parameters, and filenames through pull-down menus and pop-up dialog boxes, report messages in a scrolling text field, and display images with overlaid text and graphics; the displayed images can be updated by the program, and zoomed or panned either by the program or directly by the user through mouse actions. Although basic, these capabilities appear sufficient for many prototyping tasks in computer vision. Figure 3 shows the appearance of a typical VX-based program.

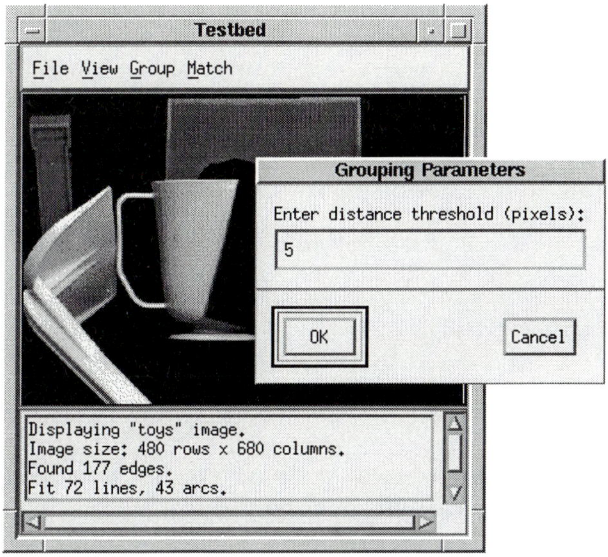

Figure 3. Appearance of a typical program constructed using Vista's VX package.

Like most programs implementing graphic user interfaces, a VX-based program has two stages. During the first stage, various calls are made to VX routines to parse program arguments and to define the commands that will appear on the menu bar. Each menu command is associated with a callback routine supplied by the programmer; callback routines can also be attached to the image display. During the second stage, the program's operation is entirely driven by the user who, by selecting menu commands or image pixels, invokes various callback routines. The callback routines, in turn, use VX routines to pop up new dialogs, solicit new parameters, issue text messages, or update the displayed image. Programs organized this way are relatively easy to construct and modify. Moreover, where a VX-based program requires some user interface feature not supported by VX, underlying facilities of the X Window System can still be accessed directly.

8. Conclusion

Vista is available as a software environment for computer vision research. While it includes support for common data types such as images and edge vectors, it is also easily extended. Existing data types can be augmented and new ones added within the framework Vista provides. The environment is modular, lightweight, and easily mastered; it is written for portability and in a standard language; and it is freely available with no significant copyright restrictions. For these reasons, computer vision algorithms and data structures implemented in the context of Vista can be readily shared.

Not all of Vista's features can be described here. It does, however, come with extensive documentation. The most recent distribution of Vista can be obtained by anonymous FTP from the `/pub/local/vista` directory of `ftp.cs.ubc.ca`. For information on the current status of Vista's ongoing development, contact the authors.

9. Acknowledgments

Vista was developed at the University of British Columbia's Laboratory for Computational Intelligence. In addition to the authors, those who made significant contributions were Ralph Horstmann, Johnny Kam, Daniel Ko, Richard Pollock, and Dan Razzell. Many others at UBC and elsewhere have participated in its testing and refinement.

10. References

1. R. M. Haralick, "Computer vision theory: The lack thereof," *Comput. Vision Graphics Image Processing* **36** (1986), pp. 372–386.
2. K. Price, "Anything you can do, I can do better (no you can't)...," *Comput. Vision Graphics Image Processing* **36** (1986), pp. 387–391.
3. J. Rasure, D. Argiro, T. Sauer, and C. Williams, "Visual language and software development environment for image processing," *Int. J. of Imaging Systems and Technology* **2** (1990), pp. 183–199.
4. A. Jacot-Descombes, M. Rupp, and T. Pun, "LaboImage: A portable window-based environment for research in image processing and analysis," in *SPIE Proc. Vol 1659: Image Processing and Interchange: Implementation and Systems* (1992), pp. 331–340.
5. M. S. Landy, Y. Cohen, and G. Sperling, "HIPS: A Unix-based image processing system," *Comput. Vision Graphics Image Processing* **25** (1984), pp. 331–347.
6. P. Hallinan, M. Nitzberg, and G. Gordon, "The HVision Package: Image Processing Software at the Harvard Robotics Lab," Harvard Univ. Robotics Lab. Tech. Rep. 88–1, Jan. 1990.
7. "KBVision," Amerinex Artificial Intelligence, Inc., Amherst, Mass. Software.
8. J. Mundy et al., "The Image Understanding Environment program," In *Proc. Computer Vision and Pattern Recognition* (1992), pp. 406–416.
9. J. Poskanzer, "PBMPLUS," Dec. 1991. Software available on Internet.

Appendix

Table 1. Operations implemented in Vista as of fall 1993.
(Each is available as a stand-alone UNIX program and as library routine, unless stated otherwise.)

viewing data	interactive programs for viewing images and edge vectors; toolkit for generating such programs; widget for displaying images
basic image processing	scale, crop, flip, transpose or rotate an image; adjust image brightness and contrast; perform an arithmetic or logical operation on each pixel of an image; convert a color image to gray-scale; compute pixel statistics
image filtering	convolve two images; convolve an image with a Gaussian kernel, or with a derivative of a Gaussian kernel
Fourier analysis	build a complex image from real and imaginary components; compute the Fourier transform of an image; compute the magnitude or phase of a complex image
edge detection	estimate image gradient; decompose gradient into magnitude and orientation; mark zero crossings; Canny's edge detector
edge organization	link edge pixels into curves; decompose curves into straight line segments
data file manipulation	combine multiple files into a single, multi-object file; combine multiple images into a single, multi-band image; select specified objects from a file
format conversion	convert images to/from Portable Graymap/Pixmap format; convert a text file to an image
printing	render an image and/or edge vectors as an Encapsulated PostScript (EPS) document; program for arranging multiple EPS documents on a page; routines for generating PostScript documents